H.264 and MPEG-4 Video Compression

H.264 and MPEG-4 Video Compression

Video Coding for Next-generation Multimedia

Iain E. G. Richardson
The Robert Gordon University, Aberdeen, UK

WILEY

To Phyllis

Contents

About the Author

Iain Richardson is a lecturer and researcher at The Robert Gordon University, Aberdeen, Scotland. He was awarded the degrees of MEng (Heriot-Watt University) and PhD (The Robert Gordon University) in 1990 and 1999 respectively. He has been actively involved in research and development of video compression systems since 1993 and is the author of over 40 journal and conference papers and two previous books. He leads the Image Communication Technology Research Group at The Robert Gordon University and advises a number of companies on video compression technology issues.

Foreword

Work on the emerging "Advanced Video Coding" standard now known as ITU-T Recommendation H.264 and as ISO/IEC 14496 (MPEG-4) Part 10 has dominated the video coding standardization community for roughly the past three years. The work has been stimulating, intense, dynamic, and all consuming for those of us most deeply involved in its design. The time has arrived to see what has been accomplished.

Although not a direct participant, Dr Richardson was able to develop a high-quality, up-to-date, introductory description and analysis of the new standard. The timeliness of this book is remarkable, as the standard itself has only just been completed.

The new H.264/AVC standard is designed to provide a technical solution appropriate for a broad range of applications, including:

- Broadcast over cable, satellite, cable modem, DSL, terrestrial.
- Interactive or serial storage on optical and magnetic storage devices, DVD, etc.
- Conversational services over ISDN, Ethernet, LAN, DSL, wireless and mobile networks, modems.
- Video-on-demand or multimedia streaming services over cable modem, DSL, ISDN, LAN, wireless networks.
- Multimedia messaging services over DSL, ISDN.

The range of bit rates and picture sizes supported by H.264/AVC is correspondingly broad, addressing video coding capabilities ranging from very low bit rate, low frame rate, "postage stamp" resolution video for mobile and dial-up devices, through to entertainment-quality standard-definition television services, HDTV, and beyond. A flexible system interface for the coded video is specified to enable the adaptation of video content for use over this full variety of network and channel-type environments. However, at the same time, the technical design is highly focused on providing the two limited goals of high coding efficiency and robustness to network environments for conventional rectangular-picture camera-view video content. Some potentially-interesting (but currently non-mainstream) features were deliberately left out (at least from the first version of the standard) because of that focus (such as support of arbitrarily-shaped video objects, some forms of bit rate scalability, 4:2:2 and 4:4:4 chroma formats, and color sampling accuracies exceeding eight bits per color component).

In the work on the new H.264/AVC standard, a number of relatively new technical developments have been adopted. For increased coding efficiency, these include improved prediction design aspects as follows:

- Variable block-size motion compensation with small block sizes,
- Quarter-sample accuracy for motion compensation,
- Motion vectors over picture boundaries,
- Multiple reference picture motion compensation,
- Decoupling of referencing order from display order,
- Decoupling of picture representation methods from the ability to use a picture for reference,
- Weighted prediction,
- Improved "skipped" and "direct" motion inference,
- Directional spatial prediction for intra coding, and
- In-the-loop deblocking filtering.

In addition to improved prediction methods, other aspects of the design were also enhanced for improved coding efficiency, including:

- Small block-size transform,
- Hierarchical block transform,
- Short word-length transform,
- Exact-match transform,
- Arithmetic entropy coding, and
- Context-adaptive entropy coding.

And for robustness to data errors/losses and flexibility for operation over a variety of network environments, some key design aspects include:

- Parameter set structure,
- NAL unit syntax structure,
- Flexible slice size,
- Flexible macroblock ordering,
- Arbitrary slice ordering,
- Redundant pictures,
- Data partitioning, and
- SP/SI synchronization switching pictures.

Prior to the H.264/AVC project, the big recent video coding activity was the MPEG-4 Part 2 (Visual) coding standard. That specification introduced a new degree of creativity and flexibility to the capabilities of the representation of digital visual content, especially with its coding of video "objects", its scalability features, extended N-bit sample precision and 4:4:4 color format capabilities, and its handling of synthetic visual scenes. It introduced a number of design variations (called "profiles" and currently numbering 19 in all) for a wide variety of applications. The H.264/AVC project (with only 3 profiles) returns to the narrower and more traditional focus on efficient compression of generic camera-shot rectangular video pictures with robustness to network losses – making no attempt to cover the ambitious breadth of MPEG-4 Visual. MPEG-4 Visual, while not quite as "hot off the press", establishes a landmark in recent technology development, and its capabilities are yet to be fully explored.

Most people first learn about a standard in publications other than the standard itself. My personal belief is that if you want to know about a standard, you should also obtain a copy of it, read it, and refer to that document alone as the ultimate authority on its content, its boundaries, and its capabilities. No tutorial or overview presentation will provide all of the insights that can be obtained from careful analysis of the standard itself.

At the same time, no standardized specification document (at least for video coding), can be a complete substitute for a good technical book on the subject. Standards specifications are written primarily to be precise, consistent, complete, and correct and not to be particularly readable. Standards tend to leave out information that is not *absolutely* necessary to comply with them. Many people find it surprising, for example, that video coding standards say almost nothing about how an encoder works or how one should be designed. In fact an encoder is essentially allowed to do anything that produces bits that can be correctly decoded, regardless of what picture quality comes out of that decoding process. People, however, can usually only understand the principles of video coding if they think from the perspective of the encoder, and nearly all textbooks (including this one) approach the subject from the encoding perspective. A good book, such as this one, will tell you why a design is the way it is and how to make use of that design, while a good standard may only tell you exactly what it is and abruptly (deliberately) stop right there.

In the case of H.264/AVC or MPEG-4 Visual, it is highly advisable for those new to the subject to read some introductory overviews such as this one, and even to get a copy of an older and simpler standard such as H.261 or MPEG-1 and try to understand that first. The principles of digital video codec design are not too complicated, and haven't really changed much over the years – but those basic principles have been wrapped in layer-upon-layer of technical enhancements to the point that the simple and straightforward concepts that lie at their core can become obscured. The entire H.261 specification was only 25 pages long, and only 17 of those pages were actually required to fully specify the technology that now lies at the heart of all subsequent video coding standards. In contrast, the H.264/AVC and MPEG-4 Visual and specifications are more than 250 and 500 pages long, respectively, with a high density of technical detail (despite completely leaving out key information such as how to encode video using their formats). They each contain areas that are difficult even for experts to fully comprehend and appreciate.

Dr Richardson's book is not a completely exhaustive treatment of the subject. However, his approach is highly informative and provides a good initial understanding of the key concepts, and his approach is conceptually superior (and in some aspects more objective) to other treatments of video coding publications. This and the remarkable timeliness of the subject matter make this book a strong contribution to the technical literature of our community.

Gary J. Sullivan

Biography of Gary J. Sullivan, PhD

Gary J. Sullivan is the chairman of the Joint Video Team (JVT) for the development of the latest international video coding standard known as H.264/AVC, which was recently completed as a joint project between the ITU-T video coding experts group (VCEG) and the ISO/IEC moving picture experts group (MPEG).

He is also the Rapporteur of Advanced Video Coding in the ITU-T, where he has led VCEG (ITU-T Q.6/SG16) for about seven years. He is also the ITU-T video liaison representative to MPEG and served as MPEG's (ISO/IEC JTC1/SC29/WG11) video chairman from March of 2001 to May of 2002.

He is currently a program manager of video standards and technologies in the eHome A/V platforms group of Microsoft Corporation. At Microsoft he designed and remains active in the extension of DirectX® Video Acceleration API/DDI feature of the Microsoft Windows® operating system platform.

Preface

With the widespread adoption of technologies such as digital television, Internet streaming video and DVD-Video, video compression has become an essential component of broadcast and entertainment media. The success of digital TV and DVD-Video is based upon the 10-year-old MPEG-2 standard, a technology that has proved its effectiveness but is now looking distinctly old-fashioned. It is clear that the time is right to replace MPEG-2 video compression with a more effective and efficient technology that can take advantage of recent progress in processing power. For some time there has been a running debate about which technology should take up MPEG-2's mantle. The leading contenders are the International Standards known as MPEG-4 Visual and H.264.

This book aims to provide a clear, practical and unbiased guide to these two standards to enable developers, engineers, researchers and students to understand and apply them effectively. Video and image compression is a complex and extensive subject and this book keeps an unapologetically limited focus, concentrating on the standards themselves (and in the case of MPEG-4 Visual, on the elements of the standard that support coding of 'real world' video material) and on video coding concepts that directly underpin the standards. The book takes an application-based approach and places particular emphasis on tools and features that are helpful in practical applications, in order to provide practical and useful assistance to developers and adopters of these standards.

I am grateful to a number of people who helped to shape the content of this book. I received many helpful comments and requests from readers of my book *Video Codec Design*. Particular thanks are due to Gary Sullivan for taking the time to provide helpful and detailed comments, corrections and advice and for kindly agreeing to write a Foreword; to Harvey Hanna (Impact Labs Inc), Yafan Zhao (The Robert Gordon University) and Aitor Garay for reading and commenting on sections of this book during its development; to members of the Joint Video Team for clarifying many of the details of H.264; to the editorial team at John Wiley & Sons (and especially to the ever-helpful, patient and supportive Kathryn Sharples); to Phyllis for her constant support; and finally to Freya and Hugh for patiently waiting for the long-promised trip to Storybook Glen!

I very much hope that you will find this book enjoyable, readable and above all useful. Further resources and links are available at my website, http://www.vcodex.com/. I always appreciate feedback, comments and suggestions from readers and you will find contact details at this website.

Iain Richardson

Glossary

4:2:0 (sampling)	Sampling method: chrominance components have half the horizontal and vertical resolution of luminance component
4:2:2 (sampling)	Sampling method: chrominance components have half the horizontal resolution of luminance component
4:4:4 (sampling)	Sampling method: chrominance components have same resolution as luminance component
arithmetic coding	Coding method to reduce redundancy
artefact	Visual distortion in an image
ASO	Arbitrary Slice Order, in which slices may be coded out of raster sequence
BAB	Binary Alpha Block, indicates the boundaries of a region (MPEG-4 Visual)
BAP	Body Animation Parameters
Block	Region of macroblock (8×8 or 4×4) for transform purposes
block matching	Motion estimation carried out on rectangular picture areas
blocking	Square or rectangular distortion areas in an image
B-picture (slice)	Coded picture (slice) predicted using bidirectional motion compensation
CABAC	Context-based Adaptive Binary Arithmetic Coding
CAE	Context-based Arithmetic Encoding
CAVLC	Context Adaptive Variable Length Coding
chrominance	Colour difference component
CIF	Common Intermediate Format, a colour image format
CODEC	*CO*der / *DEC*oder pair
colour space	Method of representing colour images
DCT	Discrete Cosine Transform
Direct prediction	A coding mode in which no motion vector is transmitted
DPCM	Differential Pulse Code Modulation
DSCQS	Double Stimulus Continuous Quality Scale, a scale and method for subjective quality measurement
DWT	Discrete Wavelet Transform

entropy coding	Coding method to reduce redundancy
error concealment	Post-processing of a decoded image to remove or reduce visible error effects
Exp-Golomb	Exponential Golomb variable length codes
FAP	Facial Animation Parameters
FBA	Face and Body Animation
FGS	Fine Granular Scalability
field	Odd- or even-numbered lines from an interlaced video sequence
flowgraph	Pictorial representation of a transform algorithm (or the algorithm itself)
FMO	Flexible Macroblock Order, in which macroblocks may be coded out of raster sequence
Full Search	A motion estimation algorithm
GMC	Global Motion Compensation, motion compensation applied to a complete coded object (MPEG-4 Visual)
GOP	Group Of Pictures, a set of coded video images
H.261	A video coding standard
H.263	A video coding standard
H.264	A video coding standard
HDTV	High Definition Television
Huffman coding	Coding method to reduce redundancy
HVS	Human Visual System, the system by which humans perceive and interpret visual images
hybrid (CODEC)	CODEC model featuring motion compensation and transform
IEC	International Electrotechnical Commission, a standards body
Inter (coding)	Coding of video frames using temporal prediction or compensation
interlaced (video)	Video data represented as a series of fields
intra (coding)	Coding of video frames without temporal prediction
I-picture (slice)	Picture (or slice) coded without reference to any other frame
ISO	International Standards Organisation, a standards body
ITU	International Telecommunication Union, a standards body
JPEG	Joint Photographic Experts Group, a committee of ISO (also an image coding standard)
JPEG2000	An image coding standard
latency	Delay through a communication system
Level	A set of conformance parameters (applied to a Profile)
loop filter	Spatial filter placed within encoding or decoding feedback loop
Macroblock	Region of frame coded as a unit (usually 16×16 pixels in the original frame)
Macroblock partition	Region of macroblock with its own motion vector (H.264)
Macroblock sub-partition	Region of macroblock with its own motion vector (H.264)
media processor	Processor with features specific to multimedia coding and processing
motion compensation	Prediction of a video frame with modelling of motion
motion estimation	Estimation of relative motion between two or more video frames

motion vector	Vector indicating a displaced block or region to be used for motion compensation
MPEG	Motion Picture Experts Group, a committee of ISO/IEC
MPEG-1	A multimedia coding standard
MPEG-2	A multimedia coding standard
MPEG-4	A multimedia coding standard
NAL	Network Abstraction Layer
objective quality	Visual quality measured by algorithm(s)
OBMC	Overlapped Block Motion Compensation
Picture (coded)	Coded (compressed) video frame
P-picture (slice)	Coded picture (or slice) using motion-compensated prediction from one reference frame
profile	A set of functional capabilities (of a video CODEC)
progressive (video)	Video data represented as a series of complete frames
PSNR	Peak Signal to Noise Ratio, an objective quality measure
QCIF	Quarter Common Intermediate Format
quantise	Reduce the precision of a scalar or vector quantity
rate control	Control of bit rate of encoded video signal
rate–distortion	Measure of CODEC performance (distortion at a range of coded bit rates)
RBSP	Raw Byte Sequence Payload
RGB	Red/Green/Blue colour space
ringing (artefacts)	'Ripple'-like artefacts around sharp edges in a decoded image
RTP	Real Time Protocol, a transport protocol for real-time data
RVLC	Reversible Variable Length Code
scalable coding	Coding a signal into a number of layers
SI slice	Intra-coded slice used for switching between coded bitstreams (H.264)
slice	A region of a coded picture
SNHC	Synthetic Natural Hybrid Coding
SP slice	Inter-coded slice used for switching between coded bitstreams (H.264)
sprite	Texture region that may be incorporated in a series of decoded frames (MPEG-4 Visual)
statistical redundancy	Redundancy due to the statistical distribution of data
studio quality	Lossless or near-lossless video quality
subjective quality	Visual quality as perceived by human observer(s)
subjective redundancy	Redundancy due to components of the data that are subjectively insignificant
sub-pixel (motion compensation)	Motion-compensated prediction from a reference area that may be formed by interpolating between integer-valued pixel positions
test model	A software model and document that describe a reference implementation of a video coding standard
Texture	Image or residual data
Tree-structured motion compensation	Motion compensation featuring a flexible hierarchy of partition sizes (H.264)

TSS	Three Step Search, a motion estimation algorithm
VCEG	Video Coding Experts Group, a committee of ITU
VCL	Video Coding Layer
video packet	Coded unit suitable for packetisation
VLC	Variable Length Code
VLD	Variable Length Decoder
VLE	Variable Length Encoder
VLSI	Very Large Scale Integrated circuit
VO	Video Object
VOP	Video Object Plane
VQEG	Video Quality Experts Group
VQEG	Video Quality Experts Group
Weighted prediction	Motion compensation in which the prediction samples from two references are scaled
YCbCr	Luminance, Blue chrominance, Red chrominance colour space
YUV	A colour space (see YCbCr)

1

Introduction

1.1 THE SCENE

Scene 1: Your avatar (a realistic 3D model with your appearance and voice) walks through a sophisticated virtual world populated by other avatars, product advertisements and video walls. On one virtual video screen is a news broadcast from your favourite channel; you want to see more about the current financial situation and so you interact with the broadcast and pull up the latest stock market figures. On another screen you call up a videoconference link with three friends. The video images of the other participants, neatly segmented from their backgrounds, are presented against yet another virtual backdrop.

Scene 2: Your new 3G vidphone rings; you flip the lid open and answer the call. The face of your friend appears on the screen and you greet each other. Each sees a small, clear image of the other on the phone's screen, without any of the obvious 'blockiness' of older-model video phones. After the call has ended, you call up a live video feed from a football match. The quality of the basic-rate stream isn't too great and you switch seamlessly to the higher-quality (but more expensive) 'premium' stream. For a brief moment the radio signal starts to break up but all you notice is a slight, temporary distortion in the video picture.

These two scenarios illustrate different visions of the next generation of multimedia applications. The first is a vision of MPEG-4 Visual: a rich, interactive on-line world bringing together synthetic, natural, video, image, 2D and 3D 'objects'. The second is a vision of H.264/AVC: highly efficient and reliable video communications, supporting two-way, 'streaming' and broadcast applications and robust to channel transmission problems. The two standards, each with their advantages and disadvantages and each with their supporters and critics, are contenders in the race to provide video compression for next-generation communication applications.

Turn on the television and surf through tens or hundreds of digital channels. Play your favourite movies on the DVD player and breathe a sigh of relief that you can throw out your antiquated VHS tapes. Tune in to a foreign TV news broadcast on the web (still just a postage-stamp video window but the choice and reliability of video streams is growing all the time). Chat to your friends and family by PC videophone. These activities are now commonplace and unremarkable, demonstrating that digital video is well on the way to becoming a ubiquitous

H.264 and MPEG-4 Video Compression: Video Coding for Next-generation Multimedia.
Iain E. G. Richardson.   © 2003 John Wiley & Sons, Ltd. ISBN: 0-470-84837-5

and essential component of the entertainment, computing, broadcasting and communications industries.

Pervasive, seamless, high-quality digital video has been the goal of companies, researchers and standards bodies over the last two decades. In some areas (for example broadcast television and consumer video storage), digital video has clearly captured the market, whilst in others (videoconferencing, video email, mobile video), market success is perhaps still too early to judge. However, there is no doubt that digital video is a globally important industry which will continue to pervade businesses, networks and homes. The continuous evolution of the digital video industry is being driven by commercial and technical forces. The commercial drive comes from the huge revenue potential of persuading consumers and businesses (a) to replace analogue technology and older digital technology with new, efficient, high-quality digital video products and (b) to adopt new communication and entertainment products that have been made possible by the move to digital video. The technical drive comes from continuing improvements in processing performance, the availability of higher-capacity storage and transmission mechanisms and research and development of video and image processing technology.

Getting digital video from its source (a camera or a stored clip) to its destination (a display) involves a chain of components or processes. Key to this chain are the processes of compression (encoding) and decompression (decoding), in which bandwidth-intensive 'raw' digital video is reduced to a manageable size for transmission or storage, then reconstructed for display. Getting the compression and decompression processes 'right' can give a significant technical and commercial edge to a product, by providing better image quality, greater reliability and/or more flexibility than competing solutions. There is therefore a keen interest in the continuing development and improvement of video compression and decompression methods and systems. The interested parties include entertainment, communication and broadcasting companies, software and hardware developers, researchers and holders of potentially lucrative patents on new compression algorithms.

The early successes in the digital video industry (notably broadcast digital television and DVD-Video) were underpinned by international standard ISO/IEC 13818 [1], popularly known as 'MPEG-2' (after the working group that developed the standard, the Moving Picture Experts Group). Anticipation of a need for better compression tools has led to the development of two further standards for video compression, known as ISO/IEC 14496 Part 2 ('MPEG-4 Visual') [2] and ITU-T Recommendation H.264/ISO/IEC 14496 Part 10 ('H.264') [3]. MPEG-4 Visual and H.264 share the same ancestry and some common features (they both draw on well-proven techniques from earlier standards) but have notably different visions, seeking to improve upon the older standards in different ways. The vision of MPEG-4 Visual is to move away from a restrictive reliance on rectangular video images and to provide an open, flexible framework for visual communications that uses the best features of efficient video compression and object-oriented processing. In contrast, H.264 has a more pragmatic vision, aiming to do what previous standards did (provide a mechanism for the compression of rectangular video images) but to do it in a more efficient, robust and practical way, supporting the types of applications that are becoming widespread in the marketplace (such as broadcast, storage and streaming).

At the present time there is a lively debate about which (if either) of these standards will come to dominate the market. MPEG-4 Visual is the more mature of the two new standards (its first Edition was published in 1999, whereas H.264 became an International

Standard/Recommendation in 2003). There is no doubt that H.264 can out-perform MPEG-4 Visual in compression efficiency but it does not have the older standard's bewildering flexibility. The licensing situation with regard to MPEG-4 Visual is clear (and not popular with some parts of the industry) but the cost of licensing H.264 remains to be agreed. This book is about these two important new standards and examines the background to the standards, the core concepts and technical details of each standard and the factors that will determine the answer to the question 'MPEG-4 Visual or H.264?'.

1.2 VIDEO COMPRESSION

Network bitrates continue to increase (dramatically in the local area and somewhat less so in the wider area), high bitrate connections to the home are commonplace and the storage capacity of hard disks, flash memories and optical media is greater than ever before. With the price per transmitted or stored bit continually falling, it is perhaps not immediately obvious why video compression is necessary (and why there is such a significant effort to make it better). Video compression has two important benefits. First, it makes it possible to use digital video in transmission and storage environments that would not support uncompressed ('raw') video. For example, current Internet throughput rates are insufficient to handle uncompressed video in real time (even at low frame rates and/or small frame size). A Digital Versatile Disk (DVD) can only store a few seconds of raw video at television-quality resolution and frame rate and so DVD-Video storage would not be practical without video and audio compression. Second, video compression enables more efficient use of transmission and storage resources. If a high bitrate transmission channel is available, then it is a more attractive proposition to send high-resolution compressed video or multiple compressed video channels than to send a single, low-resolution, uncompressed stream. Even with constant advances in storage and transmission capacity, compression is likely to be an essential component of multimedia services for many years to come.

An information-carrying signal may be compressed by removing redundancy from the signal. In a lossless compression system statistical redundancy is removed so that the original signal can be perfectly reconstructed at the receiver. Unfortunately, at the present time lossless methods can only achieve a modest amount of compression of image and video signals. Most practical video compression techniques are based on lossy compression, in which greater compression is achieved with the penalty that the decoded signal is not identical to the original. The goal of a video compression algorithm is to achieve efficient compression whilst minimising the distortion introduced by the compression process.

Video compression algorithms operate by removing redundancy in the temporal, spatial and/or frequency domains. Figure 1.1 shows an example of a single video frame. Within the highlighted regions, there is little variation in the content of the image and hence there is significant spatial redundancy. Figure 1.2 shows the same frame after the background region has been low-pass filtered (smoothed), removing some of the higher-frequency content. The human eye and brain (Human Visual System) are more sensitive to lower frequencies and so the image is still recognisable despite the fact that much of the 'information' has been removed. Figure 1.3 shows the next frame in the video sequence. The sequence was captured from a camera at 25 frames per second and so there is little change between the two frames in the short interval of 1/25 of a second. There is clearly significant temporal redundancy, i.e. most

Figure 1.1   Video frame (showing examples of homogeneous regions)

Figure 1.2   Video frame (low-pass filtered background)

Figure 1.3   Video frame 2

of the image remains unchanged between successive frames. By removing different types of redundancy (spatial, frequency and/or temporal) it is possible to compress the data significantly at the expense of a certain amount of information loss (distortion). Further compression can be achieved by encoding the processed data using an entropy coding scheme such as Huffman coding or Arithmetic coding.

Image and video compression has been a very active field of research and development for over 20 years and many different systems and algorithms for compression and decompression have been proposed and developed. In order to encourage interworking, competition and increased choice, it has been necessary to define standard methods of compression encoding and decoding to allow products from different manufacturers to communicate effectively. This has led to the development of a number of key International Standards for image and video compression, including the JPEG, MPEG and H.26× series of standards.

1.3 MPEG-4 AND H.264

MPEG-4 Visual and H.264 (also known as Advanced Video Coding) are standards for the coded representation of visual information. Each standard is a document that primarily defines two things, a coded representation (or syntax) that describes visual data in a compressed form and a method of decoding the syntax to reconstruct visual information. Each standard aims to ensure that compliant encoders and decoders can successfully interwork with each other, whilst allowing manufacturers the freedom to develop competitive and innovative products. The standards specifically do not define an encoder; rather, they define the output that an encoder should produce. A decoding method is defined in each standard but manufacturers are free to develop alternative decoders as long as they achieve the same result as the method in the standard.

MPEG-4 Visual (Part 2 of the MPEG-4 group of standards) was developed by the Moving Picture Experts Group (MPEG), a working group of the International Organisation for Standardisation (ISO). This group of several hundred technical experts (drawn from industry and research organisations) meet at 2–3 month intervals to develop the MPEG series of standards. MPEG-4 (a multi-part standard covering audio coding, systems issues and related aspects of audio/visual communication) was first conceived in 1993 and Part 2 was standardised in 1999. The H.264 standardisation effort was initiated by the Video Coding Experts Group (VCEG), a working group of the International Telecommunication Union (ITU-T) that operates in a similar way to MPEG and has been responsible for a series of visual telecommunication standards. The final stages of developing the H.264 standard have been carried out by the Joint Video Team, a collaborative effort of both VCEG and MPEG, making it possible to publish the final standard under the joint auspices of ISO/IEC (as MPEG-4 Part 10) and ITU-T (as Recommendation H.264) in 2003.

MPEG-4 Visual and H.264 have related but significantly different visions. Both are concerned with compression of visual data but MPEG-4 Visual emphasises flexibility whilst H.264's emphasis is on efficiency and reliability. MPEG-4 Visual provides a highly flexible toolkit of coding techniques and resources, making it possible to deal with a wide range of types of visual data including rectangular frames ('traditional' video material), video objects (arbitrary-shaped regions of a visual scene), still images and hybrids of natural (real-world) and synthetic (computer-generated) visual information. MPEG-4 Visual provides its functionality through a set of coding tools, organised into 'profiles', recommended groupings of tools suitable for certain applications. Classes of profiles include 'simple' profiles (coding of rectangular video frames), object-based profiles (coding of arbitrary-shaped visual objects), still texture profiles (coding of still images or 'texture'), scalable profiles (coding at multiple resolutions or quality levels) and studio profiles (coding for high-quality studio applications).

In contrast with the highly flexible approach of MPEG-4 Visual, H.264 concentrates specifically on efficient compression of video frames. Key features of the standard include compression efficiency (providing significantly better compression than any previous standard), transmission efficiency (with a number of built-in features to support reliable, robust transmission over a range of channels and networks) and a focus on popular applications of video compression. Only three profiles are currently supported (in contrast to nearly 20 in MPEG-4 Visual), each targeted at a class of popular video communication applications. The Baseline profile may be particularly useful for "conversational" applications such as video-conferencing, the Extended profile adds extra tools that are likely to be useful for video streaming across networks and the Main profile includes tools that may be suitable for consumer applications such as video broadcast and storage.

1.4 THIS BOOK

The aim of this book is to provide a technically-oriented guide to the MPEG-4 Visual and H.264/AVC standards, with an emphasis on practical issues. Other works cover the details of the other parts of the MPEG-4 standard [4–6] and this book concentrates on the application of MPEG-4 Visual and H.264 to the coding of natural video. Most practical applications of

MPEG-4 (and emerging applications of H.264) make use of a subset of the tools provided by each standard (a 'profile') and so the treatment of each standard in this book is organised according to profile, starting with the most basic profiles and then introducing the extra tools supported by more advanced profiles.

Chapters 2 and 3 cover essential background material that is required for an understanding of both MPEG-4 Visual and H.264. Chapter 2 introduces the basic concepts of digital video including capture and representation of video in digital form, colour-spaces, formats and quality measurement. Chapter 3 covers the fundamentals of video compression, concentrating on aspects of the compression process that are common to both standards and introducing the transform-based CODEC 'model' that is at the heart of all of the major video coding standards.

Chapter 4 looks at the standards themselves and examines the way that the standards have been shaped and developed, discussing the composition and procedures of the VCEG and MPEG standardisation groups. The chapter summarises the content of the standards and gives practical advice on how to approach and interpret the standards and ensure conformance. Related image and video coding standards are briefly discussed.

Chapters 5 and 6 focus on the technical features of MPEG-4 Visual and H.264. The approach is based on the structure of the Profiles of each standard (important conformance points for CODEC developers). The Simple Profile (and related Profiles) have shown themselves to be by far the most popular features of MPEG-4 Visual to date and so Chapter 5 concentrates first on the compression tools supported by these Profiles, followed by the remaining (less commercially popular) Profiles supporting coding of video objects, still texture, scalable objects and so on. Because this book is primarily about compression of natural (real-world) video information, MPEG-4 Visual's synthetic visual tools are covered only briefly. H.264's Baseline Profile is covered first in Chapter 6, followed by the extra tools included in the Main and Extended Profiles. Chapters 5 and 6 make extensive reference back to Chapter 3 (Video Coding Concepts). H.264 is dealt with in greater technical detail than MPEG-4 Visual because of the limited availability of reference material on the newer standard.

Practical issues related to the design and performance of video CODECs are discussed in Chapter 7. The design requirements of each of the main functional modules required in a practical encoder or decoder are addressed, from motion estimation through to entropy coding. The chapter examines interface requirements and practical approaches to pre- and post-processing of video to improve compression efficiency and/or visual quality. The compression and computational performance of the two standards is compared and rate control (matching the encoder output to practical transmission or storage mechanisms) and issues faced in transporting and storing of compressed video are discussed.

Chapter 8 examines the requirements of some current and emerging applications, lists some currently-available CODECs and implementation platforms and discusses the important implications of commercial factors such as patent licenses. Finally, some predictions are made about the next steps in the standardisation process and emerging research issues that may influence the development of future video coding standards.

1.5 REFERENCES

1. ISO/IEC 13818, Information Technology – Generic Coding of Moving Pictures and Associated Audio Information, 2000.

2. ISO/IEC 14496-2, Coding of Audio-Visual Objects – Part 2:Visual, 2001.
3. ISO/IEC 14496-10 and ITU-T Rec. H.264, Advanced Video Coding, 2003.
4. F. Pereira and T. Ebrahimi (eds), *The MPEG-4 Book*, IMSC Press, 2002.
5. A. Walsh and M. Bourges-Sévenier (eds), *MPEG-4 Jump Start*, Prentice-Hall, 2002.
6. ISO/IEC JTC1/SC29/WG11   N4668,   MPEG-4   Overview,   http://www.m4if.org/resources/
     Overview.pdf, March 2002.

2

Video Formats and Quality

2.1 INTRODUCTION

Video coding is the process of compressing and decompressing a digital video signal. This chapter examines the structure and characteristics of digital images and video signals and introduces concepts such as sampling formats and quality metrics that are helpful to an understanding of video coding. Digital video is a representation of a natural (real-world) visual scene, sampled spatially and temporally. A scene is sampled at a point in time to produce a frame (a representation of the complete visual scene at that point in time) or a field (consisting of odd- or even-numbered lines of spatial samples). Sampling is repeated at intervals (e.g. 1/25 or 1/30 second intervals) to produce a moving video signal. Three sets of samples (components) are typically required to represent a scene in colour. Popular formats for representing video in digital form include the ITU-R 601 standard and the set of 'intermediate formats'. The accuracy of a reproduction of a visual scene must be measured to determine the performance of a visual communication system, a notoriously difficult and inexact process. Subjective measurements are time consuming and prone to variations in the response of human viewers. Objective (automatic) measurements are easier to implement but as yet do not accurately match the opinion of a 'real' human.

2.2 NATURAL VIDEO SCENES

A typical 'real world' or 'natural' video scene is composed of multiple objects each with their own characteristic shape, depth, texture and illumination. The colour and brightness of a natural video scene changes with varying degrees of smoothness throughout the scene ('continuous tone'). Characteristics of a typical natural video scene (Figure 2.1) that are relevant for video processing and compression include spatial characteristics (texture variation within scene, number and shape of objects, colour, etc.) and temporal characteristics (object motion, changes in illumination, movement of the camera or viewpoint and so on).

H.264 and MPEG-4 Video Compression: Video Coding for Next-generation Multimedia.
Iain E. G. Richardson.   © 2003 John Wiley & Sons, Ltd. ISBN: 0-470-84837-5

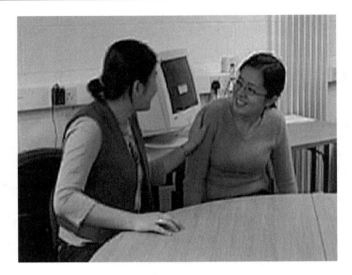

Figure 2.1   Still image from natural video scene

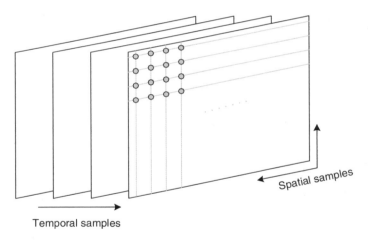

Figure 2.2   Spatial and temporal sampling of a video sequence

2.3 CAPTURE

A natural visual scene is spatially and temporally continuous. Representing a visual scene in digital form involves sampling the real scene spatially (usually on a rectangular grid in the video image plane) and temporally (as a series of still frames or components of frames sampled at regular intervals in time) (Figure 2.2). Digital video is the representation of a sampled video scene in digital form. Each spatio-temporal sample (picture element or pixel) is represented as a number or set of numbers that describes the brightness (luminance) and colour of the sample.

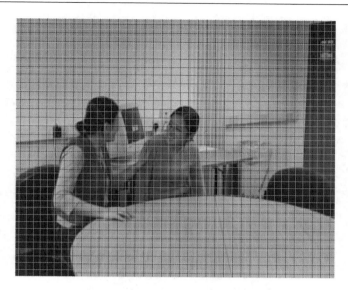

Figure 2.3   Image with 2 sampling grids

To obtain a 2D sampled image, a camera focuses a 2D projection of the video scene onto a sensor, such as an array of Charge Coupled Devices (CCD array). In the case of colour image capture, each colour component is separately filtered and projected onto a CCD array (see Section 2.4).

2.3.1 Spatial Sampling

The output of a CCD array is an analogue video signal, a varying electrical signal that represents a video image. Sampling the signal at a point in time produces a sampled image or frame that has defined values at a set of sampling points. The most common format for a sampled image is a rectangle with the sampling points positioned on a square or rectangular grid. Figure 2.3 shows a continuous-tone frame with two different sampling grids superimposed upon it. Sampling occurs at each of the intersection points on the grid and the sampled image may be reconstructed by representing each sample as a square picture element (pixel). The visual quality of the image is influenced by the number of sampling points. Choosing a 'coarse' sampling grid (the black grid in Figure 2.3) produces a low-resolution sampled image (Figure 2.4) whilst increasing the number of sampling points slightly (the grey grid in Figure 2.3) increases the resolution of the sampled image (Figure 2.5).

2.3.2 Temporal Sampling

A moving video image is captured by taking a rectangular 'snapshot' of the signal at periodic time intervals. Playing back the series of frames produces the appearance of motion. A higher temporal sampling rate (frame rate) gives apparently smoother motion in the video scene but requires more samples to be captured and stored. Frame rates below 10 frames per second are sometimes used for very low bit-rate video communications (because the amount of data

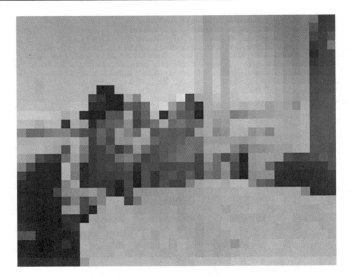

Figure 2.4   Image sampled at coarse resolution (black sampling grid)

Figure 2.5   Image sampled at slightly finer resolution (grey sampling grid)

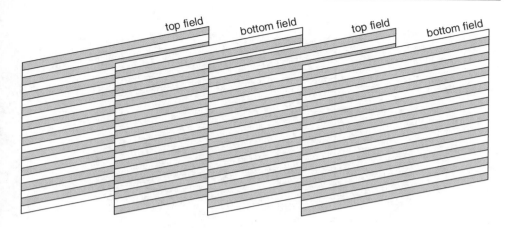

Figure 2.6 Interlaced video sequence

is relatively small) but motion is clearly jerky and unnatural at this rate. Between 10 and 20 frames per second is more typical for low bit-rate video communications; the image is smoother but jerky motion may be visible in fast-moving parts of the sequence. Sampling at 25 or 30 complete frames per second is standard for television pictures (with interlacing to improve the appearance of motion, see below); 50 or 60 frames per second produces smooth apparent motion (at the expense of a very high data rate).

2.3.3 Frames and Fields

A video signal may be sampled as a series of complete frames (*progressive* sampling) or as a sequence of interlaced fields (*interlaced* sampling). In an interlaced video sequence, half of the data in a frame (one field) is sampled at each temporal sampling interval. A field consists of either the odd-numbered or even-numbered lines within a complete video frame and an interlaced video sequence (Figure 2.6) contains a series of fields, each representing half of the information in a complete video frame (e.g. Figure 2.7 and Figure 2.8). The advantage of this sampling method is that it is possible to send twice as many fields per second as the number of frames in an equivalent progressive sequence with the same data rate, giving the appearance of smoother motion. For example, a PAL video sequence consists of 50 fields per second and, when played back, motion can appears smoother than in an equivalent progressive video sequence containing 25 frames per second.

2.4 COLOUR SPACES

Most digital video applications rely on the display of colour video and so need a mechanism to capture and represent colour information. A monochrome image (e.g. Figure 2.1) requires just one number to indicate the brightness or luminance of each spatial sample. Colour images, on the other hand, require at least three numbers per pixel position to represent colour accurately. The method chosen to represent brightness (luminance or luma) and colour is described as a colour space.

Figure 2.7 Top field

Figure 2.8 Bottom field

2.4.1 RGB

In the RGB colour space, a colour image sample is represented with three numbers that indicate the relative proportions of Red, Green and Blue (the three additive primary colours of light). Any colour can be created by combining red, green and blue in varying proportions. Figure 2.9 shows the red, green and blue components of a colour image: the red component consists of all the red samples, the green component contains all the green samples and the blue component contains the blue samples. The person on the right is wearing a blue sweater and so this appears 'brighter' in the blue component, whereas the red waistcoat of the figure on the left

Figure 2.9 Red, Green and Blue components of colour image

appears brighter in the red component. The RGB colour space is well-suited to capture and display of colour images. Capturing an RGB image involves filtering out the red, green and blue components of the scene and capturing each with a separate sensor array. Colour Cathode Ray Tubes (CRTs) and Liquid Crystal Displays (LCDs) display an RGB image by separately illuminating the red, green and blue components of each pixel according to the intensity of each component. From a normal viewing distance, the separate components merge to give the appearance of 'true' colour.

2.4.2 YCbCr

The human visual system (HVS) is less sensitive to colour than to luminance (brightness). In the RGB colour space the three colours are equally important and so are usually all stored at the same resolution but it is possible to represent a colour image more efficiently by separating the luminance from the colour information and representing luma with a higher resolution than colour.

The YCbCr colour space and its variations (sometimes referred to as YUV) is a popular way of efficiently representing colour images. Y is the luminance (luma) component and can be calculated as a weighted average of R, G and B:

$$Y = k_r R + k_g G + k_b B \qquad (2.1)$$

where k are weighting factors.

The colour information can be represented as *colour difference* (chrominance or chroma) components, where each chrominance component is the difference between R, G or B and the luminance Y:

$$Cb = B - Y$$
$$Cr = R - Y \qquad (2.2)$$
$$Cg = G - Y$$

The complete description of a colour image is given by Y (the luminance component) and three colour differences Cb, Cr and Cg that represent the difference between the colour intensity and the mean luminance of each image sample. Figure 2.10 shows the chroma components (red, green and blue) corresponding to the RGB components of Figure 2.9. Here, mid-grey is zero difference, light grey is a positive difference and dark grey is a negative difference. The chroma components only have significant values where there is a large

Figure 2.10   Cr, Cg and Cb components

difference between the colour component and the luma image (Figure 2.1). Note the strong blue and red difference components.

So far, this representation has little obvious merit since we now have four components instead of the three in RGB. However, $Cb + Cr + Cg$ is a constant and so only two of the three chroma components need to be stored or transmitted since the third component can always be calculated from the other two. In the YCbCr colour space, only the luma (Y) and blue and red chroma (Cb, Cr) are transmitted. YCbCr has an important advantage over RGB, that is the Cr and Cb components may be represented with a *lower resolution* than Y because the HVS is less sensitive to colour than luminance. This reduces the amount of data required to represent the chrominance components without having an obvious effect on visual quality. To the casual observer, there is no obvious difference between an RGB image and a YCbCr image with reduced chrominance resolution. Representing chroma with a lower resolution than luma in this way is a simple but effective form of image compression.

An RGB image may be converted to YCbCr after capture in order to reduce storage and/or transmission requirements. Before displaying the image, it is usually necessary to convert back to RGB. The equations for converting an RGB image to and from YCbCr colour space and vice versa are given in Equation 2.3 and Equation 2.4[1]. Note that there is no need to specify a separate factor k_g (because $k_b + k_r + k_g = 1$) and that G can be extracted from the YCbCr representation by subtracting Cr and Cb from Y, demonstrating that it is not necessary to store or transmit a Cg component.

$$Y = k_r R + (1 - k_b - k_r)G + k_b B$$
$$Cb = \frac{0.5}{1 - k_b}(B - Y) \tag{2.3}$$
$$Cr = \frac{0.5}{1 - k_r}(R - Y)$$

$$R = Y + \frac{1 - k_r}{0.5}Cr$$
$$G = Y - \frac{2k_b(1 - k_b)}{1 - k_b - k_r}Cb - \frac{2k_r(1 - k_r)}{1 - k_b - k_r}Cr \tag{2.4}$$
$$B = Y + \frac{1 - k_b}{0.5}Cb$$

[1] Thanks to Gary Sullivan for suggesting the form of Equations 2.3 and 2.4

ITU-R recommendation BT.601 [1] defines $k_b = 0.114$ and $k_r = 0.299$. Substituting into the above equations gives the following widely-used conversion equations:

$$Y = 0.299R + 0.587G + 0.114B$$
$$Cb = 0.564(B - Y) \tag{2.5}$$
$$Cr = 0.713(R - Y)$$

$$R = Y + 1.402Cr$$
$$G = Y - 0.344Cb - 0.714Cr \tag{2.6}$$
$$B = Y + 1.772Cb$$

2.4.3 YCbCr Sampling Formats

Figure 2.11 shows three sampling patterns for Y, Cb and Cr that are supported by MPEG-4 Visual and H.264. 4:4:4 sampling means that the three components (Y, Cb and Cr) have the same resolution and hence a sample of each component exists at every pixel position. The numbers indicate the relative sampling rate of each component in the *horizontal* direction, i.e. for every four luminance samples there are four Cb and four Cr samples. 4:4:4 sampling preserves the full fidelity of the chrominance components. In 4:2:2 sampling (sometimes referred to as YUY2), the chrominance components have the same vertical resolution as the luma but half the horizontal resolution (the numbers 4:2:2 mean that for every four luminance samples in the horizontal direction there are two Cb and two Cr samples). 4:2:2 video is used for high-quality colour reproduction.

In the popular 4:2:0 sampling format ('YV12'), Cb and Cr each have half the horizontal and vertical resolution of Y. The term '4:2:0' is rather confusing because the numbers do not actually have a logical interpretation and appear to have been chosen historically as a 'code' to identify this particular sampling pattern and to differentiate it from 4:4:4 and 4:2:2. 4:2:0 sampling is widely used for consumer applications such as video conferencing, digital television and digital versatile disk (DVD) storage. Because each colour difference component contains one quarter of the number of samples in the Y component, 4:2:0 YCbCr video requires exactly half as many samples as 4:4:4 (or R:G:B) video.

Example

Image resolution: 720 × 576 pixels
Y resolution: 720 × 576 samples, each represented with eight bits

4:4:4 Cb, Cr resolution: 720 × 576 samples, each eight bits
Total number of bits: 720 × 576 × 8 × 3 = 9 953 280 bits

4:2:0 Cb, Cr resolution: 360 × 288 samples, each eight bits
Total number of bits: (720 × 576 × 8) + (360 × 288 × 8 × 2) = 4 976 640 bits

The 4:2:0 version requires half as many bits as the 4:4:4 version.

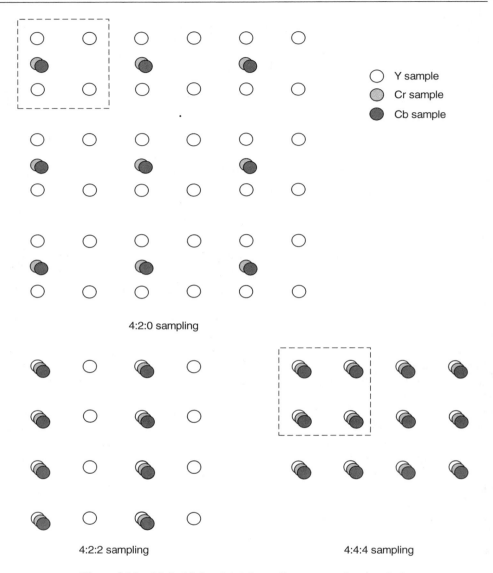

Figure 2.11   4:2:0, 4:2:2 and 4:4:4 sampling patterns (progressive)

4:2:0 sampling is sometimes described as '12 bits per pixel'. The reason for this can be seen by examining a group of four pixels (see the groups enclosed in dotted lines in Figure 2.11). Using 4:4:4 sampling, a total of 12 samples are required, four each of Y, Cb and Cr, requiring a total of $12 \times 8 = 96$ bits, an average of $96/4 = 24$ bits per pixel. Using 4:2:0 sampling, only six samples are required, four Y and one each of Cb, Cr, requiring a total of $6 \times 8 = 48$ bits, an average of $48/4 = 12$ bits per pixel.

In a 4:2:0 interlaced video sequence, the Y, Cb and Cr samples corresponding to a complete video frame are allocated to two fields. Figure 2.12 shows the method of allocating

Table 2.1   Video frame formats

Format	Luminance resolution (horiz. × vert.)	Bits per frame (4:2:0, eight bits per sample)
Sub-QCIF	128 × 96	147456
Quarter CIF (QCIF)	176 × 144	304128
CIF	352 × 288	1216512
4CIF	704 × 576	4866048

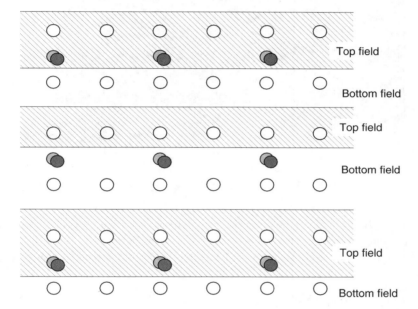

Figure 2.12   Allocaton of 4:2:0 samples to top and bottom fields

Y, Cb and Cr samples to a pair of interlaced fields adopted in MPEG-4 Visual and H.264. It is clear from this figure that the total number of samples in a pair of fields is the same as the number of samples in an equivalent progressive frame.

2.5 VIDEO FORMATS

The video compression standards described in this book can compress a wide variety of video frame formats. In practice, it is common to capture or convert to one of a set of 'intermediate formats' prior to compression and transmission. The Common Intermediate Format (CIF) is the basis for a popular set of formats listed in Table 2.1. Figure 2.13 shows the luma component of a video frame sampled at a range of resolutions, from 4CIF down to Sub-QCIF. The choice of frame resolution depends on the application and available storage or transmission capacity. For example, 4CIF is appropriate for standard-definition television and DVD-video; CIF and QCIF

Figure 2.13   Video frame sampled at range of resolutions

are popular for videoconferencing applications; QCIF or SQCIF are appropriate for mobile multimedia applications where the display resolution and the bitrate are limited. Table 2.1 lists the number of bits required to represent one uncompressed frame in each format (assuming 4:2:0 sampling and 8 bits per luma and chroma sample).

A widely-used format for digitally coding video signals for television production is ITU-R Recommendation BT.601-5 [1] (the term 'coding' in the Recommendation title means conversion to digital format and does not imply compression). The luminance component of the video signal is sampled at 13.5 MHz and the chrominance at 6.75 MHz to produce a 4:2:2 Y:Cb:Cr component signal. The parameters of the sampled digital signal depend on the video frame rate (30 Hz for an NTSC signal and 25 Hz for a PAL/SECAM signal) and are shown in Table 2.2. The higher 30 Hz frame rate of NTSC is compensated for by a lower spatial resolution so that the total bit rate is the same in each case (216 Mbps). The actual area shown on the display, the *active area*, is smaller than the total because it excludes horizontal and vertical blanking intervals that exist 'outside' the edges of the frame.

Each sample has a possible range of 0 to 255. Levels of 0 and 255 are reserved for synchronisation and the active luminance signal is restricted to a range of 16 (black) to 235 (white).

2.6 QUALITY

In order to specify, evaluate and compare video communication systems it is necessary to determine the quality of the video images displayed to the viewer. Measuring visual quality is

Table 2.2 ITU-R BT.601-5 Parameters

	30 Hz frame rate	25 Hz frame rate
Fields per second	60	50
Lines per complete frame	525	625
Luminance samples per line	858	864
Chrominance samples per line	429	432
Bits per sample	8	8
Total bit rate	216 Mbps	216 Mbps
Active lines per frame	480	576
Active samples per line (Y)	720	720
Active samples per line (Cr,Cb)	360	360

a difficult and often imprecise art because there are so many factors that can affect the results. Visual quality is inherently *subjective* and is influenced by many factors that make it difficult to obtain a completely accurate measure of quality. For example, a viewer's opinion of visual quality can depend very much on the task at hand, such as passively watching a DVD movie, actively participating in a videoconference, communicating using sign language or trying to identify a person in a surveillance video scene. Measuring visual quality using *objective* criteria gives accurate, repeatable results but as yet there are no objective measurement systems that completely reproduce the subjective experience of a human observer watching a video display.

2.6.1 Subjective Quality Measurement

2.6.1.1 Factors Influencing Subjective Quality

Our perception of a visual scene is formed by a complex interaction between the components of the Human Visual System (HVS), the eye and the brain. The perception of visual quality is influenced by spatial fidelity (how clearly parts of the scene can be seen, whether there is any obvious distortion) and temporal fidelity (whether motion appears natural and 'smooth'). However, a viewer's opinion of 'quality' is also affected by other factors such as the viewing environment, the observer's state of mind and the extent to which the observer interacts with the visual scene. A user carrying out a specific task that requires concentration on part of a visual scene will have a quite different requirement for 'good' quality than a user who is passively watching a movie. For example, it has been shown that a viewer's opinion of visual quality is measurably higher if the viewing environment is comfortable and non-distracting (regardless of the 'quality' of the visual image itself).

Other important influences on perceived quality include visual attention (an observer perceives a scene by fixating on a sequence of points in the image rather than by taking in everything simultaneously) and the so-called 'recency effect' (our opinion of a visual sequence is more heavily influenced by recently-viewed material than older video material) [2, 3]. All of these factors make it very difficult to measure visual quality accurately and quantitavely.

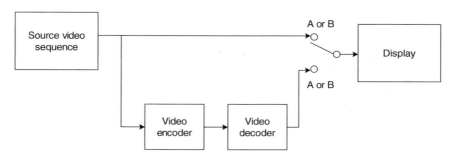

Figure 2.14   DSCQS testing system

2.6.1.2 ITU-R 500

Several test procedures for subjective quality evaluation are defined in ITU-R Recommendation BT.500-11 [4]. A commonly-used procedure from the standard is the Double Stimulus Continuous Quality Scale (DSCQS) method in which an assessor is presented with a pair of images or short video sequences A and B, one after the other, and is asked to give A and B a 'quality score' by marking on a continuous line with five intervals ranging from 'Excellent' to 'Bad'. In a typical test session, the assessor is shown a series of pairs of sequences and is asked to grade each pair. Within each pair of sequences, one is an unimpaired "reference" sequence and the other is the same sequence, modified by a system or process under test. Figure 2.14 shows an experimental set-up appropriate for the testing of a video CODEC in which the original sequence is compared with the same sequence after encoding and decoding. The selection of which sequence is 'A' and which is 'B' is randomised.

The order of the two sequences, original and "impaired", is randomised during the test session so that the assessor does not know which is the original and which is the impaired sequence. This helps prevent the assessor from pre-judging the impaired sequence compared with the reference sequence. At the end of the session, the scores are converted to a normalised range and the end result is a score (sometimes described as a 'mean opinion score') that indicates the *relative* quality of the impaired and reference sequences.

Tests such as DSCQS are accepted to be realistic measures of subjective visual quality. However, this type of test suffers from practical problems. The results can vary significantly depending on the assessor and the video sequence under test. This variation is compensated for by repeating the test with several sequences and several assessors. An 'expert' assessor (one who is familiar with the nature of video compression distortions or 'artefacts') may give a biased score and it is preferable to use 'nonexpert' assessors. This means that a large pool of assessors is required because a nonexpert assessor will quickly learn to recognise characteristic artefacts in the video sequences (and so become 'expert'). These factors make it expensive and time consuming to carry out the DSCQS tests thoroughly.

2.6.2 Objective Quality Measurement

The complexity and cost of subjective quality measurement make it attractive to be able to measure quality automatically using an algorithm. Developers of video compression and video

Figure 2.15 PSNR examples: (a) original; (b) 30.6 dB; (c) 28.3 dB

Figure 2.16 Image with blurred background (PSNR = 27.7 dB)

processing systems rely heavily on so-called objective (algorithmic) quality measures. The most widely used measure is Peak Signal to Noise Ratio (PSNR) but the limitations of this metric have led to many efforts to develop more sophisticated measures that approximate the response of 'real' human observers.

2.6.2.1 PSNR

Peak Signal to Noise Ratio (PSNR) (Equation 2.7) is measured on a logarithmic scale and depends on the mean squared error (MSE) of between an original and an impaired image or video frame, relative to $(2^n - 1)^2$ (the square of the highest-possible signal value in the image, where n is the number of bits per image sample).

$$PSNR_{dB} = 10 \log_{10} \frac{(2^n - 1)^2}{MSE} \qquad (2.7)$$

PSNR can be calculated easily and quickly and is therefore a very popular quality measure, widely used to compare the 'quality' of compressed and decompressed video images. Figure 2.15 shows a close-up of 3 images: the first image (a) is the original and (b) and (c) are degraded (blurred) versions of the original image. Image (b) has a measured PSNR of 30.6 dB whilst image (c) has a PSNR of 28.3 dB (reflecting the poorer image quality).

The PSNR measure suffers from a number of limitations. PSNR requires an unimpaired original image for comparison but this may not be available in every case and it may not be easy to verify that an 'original' image has perfect fidelity. PSNR does not correlate well with subjective video quality measures such as those defined in ITU-R 500. For a given image or image sequence, high PSNR usually indicates high quality and low PSNR usually indicates low quality. However, a particular value of PSNR does not necessarily equate to an 'absolute' subjective quality. For example, Figure 2.16 shows a distorted version of the original image from Figure 2.15 in which only the background of the image has been blurred. This image has a PSNR of 27.7 dB relative to the original. Most viewers would rate this image as significantly better than image (c) in Figure 2.15 because the face is clearer, contradicting the PSNR rating. This example shows that PSNR ratings do not necessarily correlate with 'true' subjective quality. In this case, a human observer gives a higher importance to the face region and so is particularly sensitive to distortion in this area.

2.6.2.2 Other Objective Quality Metrics

Because of the limitations of crude metrics such as PSNR, there has been a lot of work in recent years to try to develop a more sophisticated objective test that more closely approaches subjective test results. Many different approaches have been proposed [5, 6, 7] but none of these has emerged as a clear alternative to subjective tests. As yet there is no standardised, accurate system for objective ('automatic') quality measurement that is suitable for digitally coded video. In recognition of this, the ITU-T Video Quality Experts Group (VQEG) aim to develop standards for objective video quality evaluation [8]. The first step in this process was to test and compare potential models for objective evaluation. In March 2000, VQEG reported on the first round of tests in which ten competing systems were tested under identical conditions. Unfortunately, none of the ten proposals was considered suitable for standardisation and VQEG are completing a second round of evaluations in 2003. Unless there is a significant breakthrough in automatic quality assessment, the problem of accurate objective quality measurement is likely to remain for some time to come.

2.7 CONCLUSIONS

Sampling analogue video produces a digital video signal, which has the advantages of accuracy, quality and compatibility with digital media and transmission but which typically occupies a prohibitively large bitrate. Issues inherent in digital video systems include spatial and temporal resolution, colour representation and the measurement of visual quality. The next chapter introduces the basic concepts of video compression, necessary to accommodate digital video signals on practical storage and transmission media.

2.8 REFERENCES

1. Recommendation ITU-R BT.601-5, Studio encoding parameters of digital television for standard 4:3 and wide-screen 16:9 aspect ratios, ITU-T, 1995.

2. N. Wade and M. Swanston, *Visual Perception: An Introduction,* 2nd edition, Psychology Press, London, 2001.

3. R. Aldridge, J. Davidoff, D. Hands, M. Ghanbari and D. E. Pearson, Recency effect in the subjective assessment of digitally coded television pictures, *Proc. Fifth International Conference on Image Processing and its Applications*, Heriot-Watt University, Edinburgh, UK, July 1995.

4. Recommendation ITU-T BT.500-11, Methodology for the subjective assessment of the quality of television pictures, ITU-T, 2002.

5. C. J. van den Branden Lambrecht and O. Verscheure, Perceptual quality measure using a spatio-temporal model of the Human Visual System, Digital Video Compression Algorithms and Technologies, *Proc. SPIE*, **2668**, San Jose, 1996.

6. H. Wu, Z. Yu, S. Winkler and T. Chen, Impairment metrics for MC/DPCM/DCT encoded digital video, *Proc. PCS01*, Seoul, April 2001.

7. K. T. Tan and M. Ghanbari, A multi-metric objective picture quality measurement model for MPEG video, *IEEE Trans. Circuits and Systems for Video Technology*, **10** (7), October 2000.

8. http://www.vqeg.org/ (Video Quality Experts Group).

3

Video Coding Concepts

3.1 INTRODUCTION

> *compress vb.*: to squeeze together or compact into less space; condense

> *compress noun*: the act of compression or the condition of being compressed

Compression is the process of compacting data into a smaller number of bits. Video compression (video coding) is the process of compacting or condensing a digital video sequence into a smaller number of bits. 'Raw' or uncompressed digital video typically requires a large bitrate (approximately 216 Mbits for 1 second of uncompressed TV-quality video, see Chapter 2) and compression is necessary for practical storage and transmission of digital video.

Compression involves a complementary pair of systems, a compressor (encoder) and a decompressor (decoder). The encoder converts the source data into a compressed form (occupying a reduced number of bits) prior to transmission or storage and the decoder converts the compressed form back into a representation of the original video data. The encoder/decoder pair is often described as a *CODEC* (en*CO*der/ *DEC*oder) (Figure 3.1).

Data compression is achieved by removing *redundancy*, i.e. components that are not necessary for faithful reproduction of the data. Many types of data contain *statistical* redundancy and can be effectively compressed using *lossless* compression, so that the reconstructed data at the output of the decoder is a perfect copy of the original data. Unfortunately, lossless compression of image and video information gives only a moderate amount of compression. The best that can be achieved with current lossless image compression standards such as JPEG-LS [1] is a compression ratio of around 3–4 times. *Lossy* compression is necessary to achieve higher compression. In a lossy compression system, the decompressed data is not identical to the source data and much higher compression ratios can be achieved at the expense of a loss of visual quality. Lossy video compression systems are based on the principle of removing *subjective* redundancy, elements of the image or video sequence that can be removed without significantly affecting the viewer's perception of visual quality.

H.264 and MPEG-4 Video Compression: Video Coding for Next-generation Multimedia.
Iain E. G. Richardson.   © 2003 John Wiley & Sons, Ltd. ISBN: 0-470-84837-5

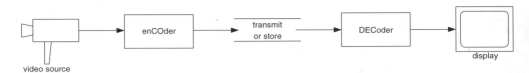

video source

Figure 3.1   Encoder/decoder

spatial correlation

Figure 3.2   Spatial and temporal correlation in a video sequence

Most video coding methods exploit both *temporal* and *spatial* redundancy to achieve compression. In the temporal domain, there is usually a high correlation (similarity) between frames of video that were captured at around the same time. Temporally adjacent frames (successive frames in time order) are often highly correlated, especially if the temporal sampling rate (the frame rate) is high. In the spatial domain, there is usually a high correlation between pixels (samples) that are close to each other, i.e. the values of neighbouring samples are often very similar (Figure 3.2).

The H.264 and MPEG-4 Visual standards (described in detail in Chapters 5 and 6) share a number of common features. Both standards assume a CODEC 'model' that uses block-based motion compensation, transform, quantisation and entropy coding. In this chapter we examine the main components of this model, starting with the temporal model (motion estimation and compensation) and continuing with image transforms, quantisation, predictive coding and entropy coding. The chapter concludes with a 'walk-through' of the basic model, following through the process of encoding and decoding a block of image samples.

3.2 VIDEO CODEC

A video CODEC (Figure 3.3) encodes a source image or video sequence into a compressed form and decodes this to produce a copy or approximation of the source sequence. If the

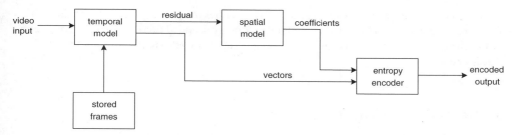

Figure 3.3   Video encoder block diagram

decoded video sequence is identical to the original, then the coding process is lossless; if the decoded sequence differs from the original, the process is lossy.

The CODEC represents the original video sequence by a *model* (an efficient coded representation that can be used to reconstruct an approximation of the video data). Ideally, the model should represent the sequence using as few bits as possible and with as high a fidelity as possible. These two goals (compression efficiency and high quality) are usually conflicting, because a lower compressed bit rate typically produces reduced image quality at the decoder. This tradeoff between bit rate and quality (the rate-distortion trade off) is discussed further in Chapter 7.

A video encoder (Figure 3.3) consists of three main functional units: a *temporal model*, a *spatial model* and an *entropy encoder*. The input to the temporal model is an uncompressed video sequence. The temporal model attempts to reduce temporal redundancy by exploiting the similarities between neighbouring video frames, usually by constructing a prediction of the current video frame. In MPEG-4 Visual and H.264, the prediction is formed from one or more previous or future frames and is improved by compensating for differences between the frames (motion compensated prediction). The output of the temporal model is a residual frame (created by subtracting the prediction from the actual current frame) and a set of model parameters, typically a set of motion vectors describing how the motion was compensated.

The residual frame forms the input to the spatial model which makes use of similarities between neighbouring samples in the residual frame to reduce spatial redundancy. In MPEG-4 Visual and H.264 this is achieved by applying a transform to the residual samples and quantizing the results. The transform converts the samples into another domain in which they are represented by transform coefficients. The coefficients are quantised to remove insignificant values, leaving a small number of significant coefficients that provide a more compact representation of the residual frame. The output of the spatial model is a set of quantised transform coefficients.

The parameters of the temporal model (typically motion vectors) and the spatial model (coefficients) are compressed by the entropy encoder. This removes statistical redundancy in the data (for example, representing commonly-occurring vectors and coefficients by short binary codes) and produces a compressed bit stream or file that may be transmitted and/or stored. A compressed sequence consists of coded motion vector parameters, coded residual coefficients and header information.

The video decoder reconstructs a video frame from the compressed bit stream. The coefficients and motion vectors are decoded by an entropy decoder after which the spatial model is decoded to reconstruct a version of the residual frame. The decoder uses the motion vector parameters, together with one or more previously decoded frames, to create a prediction of the current frame and the frame itself is reconstructed by adding the residual frame to this prediction.

3.3 TEMPORAL MODEL

The goal of the temporal model is to reduce redundancy between transmitted frames by forming a predicted frame and subtracting this from the current frame. The output of this process is a residual (difference) frame and the more accurate the prediction process, the less energy is contained in the residual frame. The residual frame is encoded and sent to the decoder which re-creates the predicted frame, adds the decoded residual and reconstructs the current frame. The predicted frame is created from one or more past or future frames ('reference frames'). The accuracy of the prediction can usually be improved by compensating for motion between the reference frame(s) and the current frame.

3.3.1 Prediction from the Previous Video Frame

The simplest method of temporal prediction is to use the previous frame as the predictor for the current frame. Two successive frames from a video sequence are shown in Figure 3.4 and Figure 3.5. Frame 1 is used as a predictor for frame 2 and the residual formed by subtracting the predictor (frame 1) from the current frame (frame 2) is shown in Figure 3.6. In this image, mid-grey represents a difference of zero and light or dark greys correspond to positive and negative differences respectively. The obvious problem with this simple prediction is that a lot of energy remains in the residual frame (indicated by the light and dark areas) and this means that there is still a significant amount of information to compress after temporal prediction. Much of the residual energy is due to object movements between the two frames and a better prediction may be formed by *compensating* for motion between the two frames.

3.3.2 Changes due to Motion

Changes between video frames may be caused by object motion (rigid object motion, for example a moving car, and deformable object motion, for example a moving arm), camera motion (panning, tilt, zoom, rotation), uncovered regions (for example, a portion of the scene background uncovered by a moving object) and lighting changes. With the exception of uncovered regions and lighting changes, these differences correspond to pixel movements between frames. It is possible to estimate the trajectory of each pixel between successive video frames, producing a field of pixel trajectories known as the *optical flow* (optic flow) [2]. Figure 3.7 shows the optical flow field for the frames of Figure 3.4 and Figure 3.5. The complete field contains a flow vector for every pixel position but for clarity, the field is sub-sampled so that only the vector for every 2nd pixel is shown.

If the optical flow field is accurately known, it should be possible to form an accurate prediction of most of the pixels of the current frame by moving each pixel from the

Figure 3.4 Frame 1

Figure 3.5 Frame 2

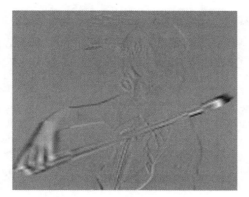

Figure 3.6 Difference

reference frame along its optical flow vector. However, this is not a practical method of motion compensation for several reasons. An accurate calculation of optical flow is very computationally intensive (the more accurate methods use an iterative procedure for every pixel) and it would be necessary to send the optical flow vector for every pixel to the decoder

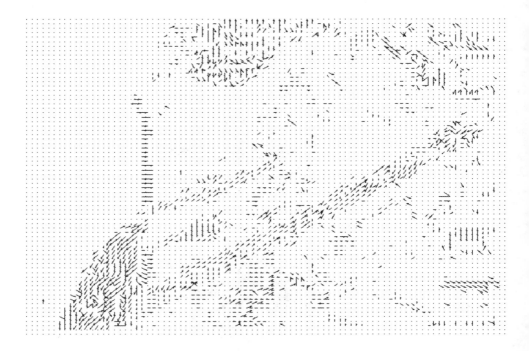

Figure 3.7 Optical flow

in order for the decoder to re-create the prediction frame (resulting in a large amount of transmitted data and negating the advantage of a small residual).

3.3.3 Block-based Motion Estimation and Compensation

A practical and widely-used method of motion compensation is to compensate for movement of rectangular sections or 'blocks' of the current frame. The following procedure is carried out for each block of $M \times N$ samples in the current frame:

1. Search an area in the reference frame (past or future frame, previously coded and transmitted) to find a 'matching' $M \times N$-sample region. This is carried out by comparing the
   $M \times N$ block in the current frame with some or all of the possible $M \times N$ regions in the
   search area (usually a region centred on the current block position) and finding the region
   that gives the 'best' match. A popular matching criterion is the energy in the residual formed
   by subtracting the candidate region from the current $M \times N$ block, so that the candidate
   region that minimises the residual energy is chosen as the best match. This process of
   finding the best match is known as *motion estimation*.
2. The chosen candidate region becomes the predictor for the current $M \times N$ block and is
   subtracted from the current block to form a residual $M \times N$ block (*motion compensation*).
3. The residual block is encoded and transmitted and the offset between the current block and
   the position of the candidate region (*motion vector*) is also transmitted.

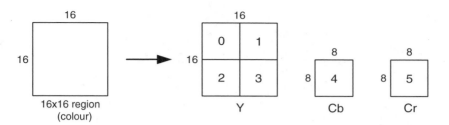

Figure 3.8   Macroblock (4:2:0)

The decoder uses the received motion vector to re-create the predictor region and decodes the residual block, adds it to the predictor and reconstructs a version of the original block.

Block-based motion compensation is popular for a number of reasons. It is relatively straightforward and computationally tractable, it fits well with rectangular video frames and with block-based image transforms (e.g. the Discrete Cosine Transform, see later) and it provides a reasonably effective temporal model for many video sequences. There are however a number of disadvantages, for example 'real' objects rarely have neat edges that match rectangular boundaries, objects often move by a fractional number of pixel positions between frames and many types of object motion are hard to compensate for using block-based methods (e.g. deformable objects, rotation and warping, complex motion such as a cloud of smoke). Despite these disadvantages, block-based motion compensation is the basis of the temporal model used by all current video coding standards.

3.3.4 Motion Compensated Prediction of a Macroblock

The *macroblock*, corresponding to a 16×16-pixel region of a frame, is the basic unit for motion compensated prediction in a number of important visual coding standards including MPEG-1, MPEG-2, MPEG-4 Visual, H.261, H.263 and H.264. For source video material in 4:2:0 format (see Chapter 2), a macroblock is organised as shown in Figure 3.8. A 16×16-pixel region of the source frame is represented by 256 luminance samples (arranged in four 8×8-sample blocks), 64 blue chrominance samples (one 8×8 block) and 64 red chrominance samples (8×8), giving a total of six 8×8 blocks. An MPEG-4 Visual or H.264 CODEC processes each video frame in units of a macroblock.

Motion Estimation

Motion estimation of a macroblock involves finding a 16×16-sample region in a reference frame that closely matches the current macroblock. The reference frame is a previously-encoded frame from the sequence and may be before or after the current frame in display order. An area in the reference frame centred on the current macroblock position (the search area) is searched and the 16×16 region within the search area that minimises a matching criterion is chosen as the 'best match' (Figure 3.9).

Motion Compensation

The selected 'best' matching region in the reference frame is subtracted from the current macroblock to produce a residual macroblock (luminance and chrominance) that is encoded

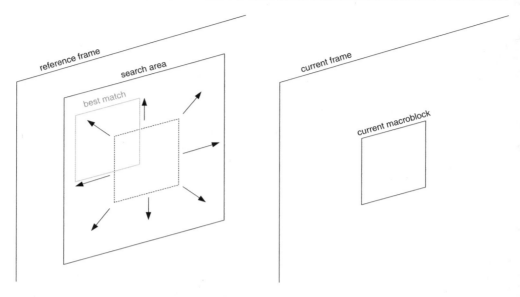

Figure 3.9   Motion estimation

and transmitted together with a motion vector describing the position of the best matching region (relative to the current macroblock position). Within the encoder, the residual is encoded and decoded and added to the matching region to form a reconstructed macroblock which is stored as a reference for further motion-compensated prediction. It is necessary to use a decoded residual to reconstruct the macroblock in order to ensure that encoder and decoder use an identical reference frame for motion compensation.

There are many variations on the basic motion estimation and compensation process. The reference frame may be a previous frame (in temporal order), a future frame or a combination of predictions from two or more previously encoded frames. If a future frame is chosen as the reference, it is necessary to encode this frame before the current frame (i.e. frames must be encoded out of order). Where there is a significant change between the reference and current frames (for example, a scene change), it may be more efficient to encode the macroblock without motion compensation and so an encoder may choose *intra* mode (encoding without motion compensation) or *inter* mode (encoding with motion compensated prediction) for each macroblock. Moving objects in a video scene rarely follow 'neat' 16×16-pixel boundaries and so it may be more efficient to use a variable block size for motion estimation and compensation. Objects may move by a fractional number of pixels between frames (e.g. 2.78 pixels rather than 2.0 pixels in the horizontal direction) and a better prediction may be formed by interpolating the reference frame to sub-pixel positions before searching these positions for the best match.

3.3.5 Motion Compensation Block Size

Two successive frames of a video sequence are shown in Figure 3.10 and Figure 3.11. Frame 1 is subtracted from frame 2 without motion compensation to produce a residual frame

Figure 3.10 Frame 1

Figure 3.11 Frame 2

Figure 3.12 Residual (no motion compensation)

Figure 3.13   Residual (16 × 16 block size)

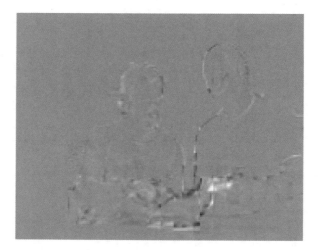

Figure 3.14   Residual (8 × 8 block size)

(Figure 3. 12). The energy in the residual is reduced by motion compensating each 16 × 16 macroblock (Figure 3.13). Motion compensating each 8 × 8 block (instead of each 16 × 16 macroblock) reduces the residual energy further (Figure 3.14) and motion compensating each 4 × 4 block gives the smallest residual energy of all (Figure 3.15). These examples show that smaller motion compensation block sizes can produce better motion compensation results. However, a smaller block size leads to increased complexity (more search operations must be carried out) and an increase in the number of motion vectors that need to be transmitted. Sending each motion vector requires bits to be sent and the extra overhead for vectors may outweigh the benefit of reduced residual energy. An effective compromise is to adapt the block size to the picture characteristics, for example choosing a large block size in flat, homogeneous regions of a frame and choosing a small block size around areas of high detail and complex

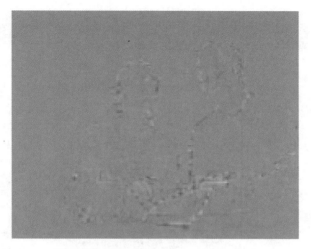

Figure 3.15 Residual (4 × 4 block size)

motion. H.264 uses an adaptive motion compensation block size (Tree Structured motion compensation, described in Chapter 6.

3.3.6 Sub-pixel Motion Compensation

Figure 3.16 shows a close-up view of part of a reference frame. In some cases a better motion compensated prediction may be formed by predicting from interpolated sample positions in the reference frame. In Figure 3.17, the reference region luma samples are interpolated to half-samples positions and it may be possible to find a better match for the current macroblock by searching the interpolated samples. 'Sub-pixel' motion estimation and compensation[1] involves searching sub-sample interpolated positions as well as integer-sample positions, choosing the position that gives the best match (i.e. minimises the residual energy) and using the integer- or sub-sample values at this position for motion compensated prediction. Figure 3.18 shows the concept of a 'quarter-pixel' motion estimation. In the first stage, motion estimation finds the best match on the integer sample grid (circles). The encoder searches the half-sample positions immediately next to this best match (squares) to see whether the match can be improved and if required, the quarter-sample positions next to the best half-sample position (triangles) are then searched. The final match (at an integer, half- or quarter-sample position) is subtracted from the current block or macroblock.

The residual in Figure 3.19 is produced using a block size of 4 × 4 samples using half-sample interpolation and has lower residual energy than Figure 3.15. This approach may be extended further by interpolation onto a quarter-sample grid to give a still smaller residual (Figure 3.20). In general, 'finer' interpolation provides better motion compensation performance (a smaller residual) at the expense of increased complexity. The performance gain tends to diminish as the interpolation steps increase. Half-sample interpolation gives a significant gain over integer-sample motion compensation, quarter-sample interpolation gives a moderate further improvement, eighth-sample interpolation gives a small further improvement again and so on.

---

[1] The terms 'sub-pixel', 'half-pixel' and 'quarter-pixel' are widely used in this context although in fact the process is usually applied to luma and chroma samples, not pixels.

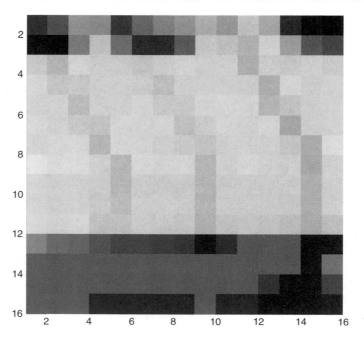

Figure 3.16   Close-up of reference region

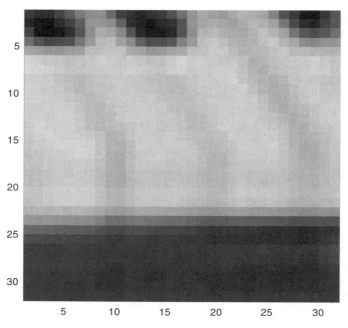

Figure 3.17   Reference region interpolated to half-pixel positions

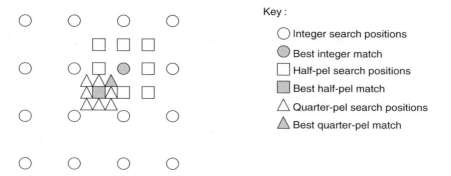

Figure 3.18   Integer, half-pixel and quarter-pixel motion estimation

Figure 3.19   Residual (4 × 4 blocks, half-pixel compensation)

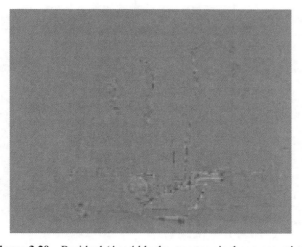

Figure 3.20   Residual (4 × 4 blocks, quarter-pixel compensation)

Table 3.1   SAE of residual frame after motion compensation (16 × 16 block size)

Sequence	No motion compensation	Integer-pel	Half-pel	Quarter-pel
'Violin', QCIF	171945	153475	128320	113744
'Grasses', QCIF	248316	245784	228952	215585
'Carphone', QCIF	102418	73952	56492	47780

Figure 3.21   Motion vector map (16 × 16 blocks, integer vectors)

Some examples of the performance achieved by sub-pixel motion estimation and compensation are given in Table 3.1. A motion-compensated reference frame (the previous frame in the sequence) is subtracted from the current frame and the energy of the residual (approximated by the Sum of Absolute Errors, SAE) is listed in the table. A lower SAE indicates better motion compensation performance. In each case, sub-pixel motion compensation gives improved performance compared with integer-sample compensation. The improvement from integer to half-sample is more significant than the further improvement from half- to quarter-sample. The sequence 'Grasses' has highly complex motion and is particularly difficult to motion-compensate, hence the large SAE; 'Violin' and 'Carphone' are less complex and motion compensation produces smaller SAE values.

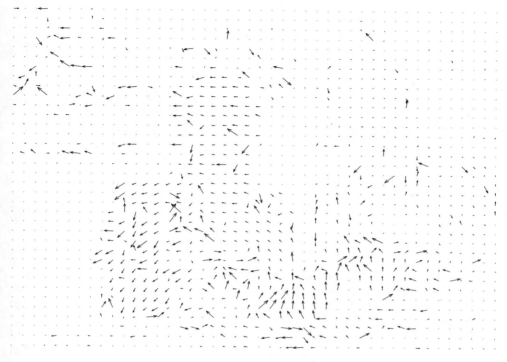

Figure 3.22 Motion vector map (4 × 4 blocks, quarter-pixel vectors)

Searching for matching 4 × 4 blocks with quarter-sample interpolation is considerably more complex than searching for 16 × 16 blocks with no interpolation. In addition to the extra complexity, there is a coding penalty since the vector for every block must be encoded and transmitted to the receiver in order to reconstruct the image correctly. As the block size is reduced, the number of vectors that have to be transmitted increases. More bits are required to represent half- or quarter-sample vectors because the fractional part of the vector (e.g. 0.25, 0.5) must be encoded as well as the integer part. Figure 3.21 plots the integer motion vectors that are required to be transmitted along with the residual of Figure 3.13. The motion vectors required for the residual of Figure 3.20 (4 × 4 block size) are plotted in Figure 3.22, in which there are 16 times as many vectors, each represented by two fractional numbers DX and DY with quarter-pixel accuracy. There is therefore a tradeoff in compression efficiency associated with more complex motion compensation schemes, since more accurate motion compensation requires more bits to encode the vector field but fewer bits to encode the residual whereas less accurate motion compensation requires fewer bits for the vector field but more bits for the residual.

3.3.7 Region-based Motion Compensation

Moving objects in a 'natural' video scene are rarely aligned neatly along block boundaries but are likely to be irregular shaped, to be located at arbitrary positions and (in some cases) to change shape between frames. This problem is illustrated by Figure 3.23, in which the

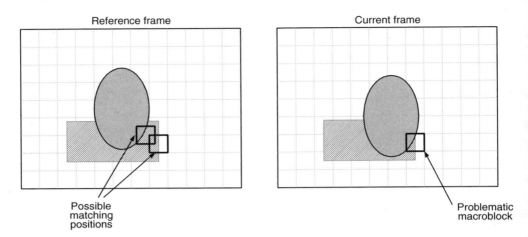

Figure 3.23   Motion compensation of arbitrary-shaped moving objects

oval-shaped object is moving and the rectangular object is static. It is difficult to find a good match in the reference frame for the highlighted macroblock, because it covers part of the moving object and part of the static object. Neither of the two matching positions shown in the reference frame are ideal.

It may be possible to achieve better performance by motion compensating arbitrary regions of the picture (region-based motion compensation). For example, if we only attempt to motion-compensate pixel positions *inside* the oval object then we can find a good match in the reference frame. There are however a number of practical difficulties that need to be overcome in order to use region-based motion compensation, including identifying the region boundaries accurately and consistently, (*segmentation*) signalling (encoding) the contour of the boundary to the decoder and encoding the residual after motion compensation. MPEG-4 Visual includes a number of tools that support region-based compensation and coding and these are described in Chapter 5.

3.4 IMAGE MODEL

A natural video image consists of a grid of sample values. Natural images are often difficult to compress in their original form because of the high correlation between neighbouring image samples. Figure 3.24 shows the two-dimensional autocorrelation function of a natural video image (Figure 3.4) in which the height of the graph at each position indicates the similarity between the original image and a spatially-shifted copy of itself. The peak at the centre of the figure corresponds to zero shift. As the spatially-shifted copy is moved away from the original image in any direction, the function drops off as shown in the figure, with the gradual slope indicating that image samples within a local neighbourhood are highly correlated.

A motion-compensated residual image such as Figure 3.20 has an autocorrelation function (Figure 3.25) that drops off rapidly as the spatial shift increases, indicating that neighbouring samples are weakly correlated. Efficient motion compensation reduces local correlation in the residual making it easier to compress than the original video frame. The function of the *image*

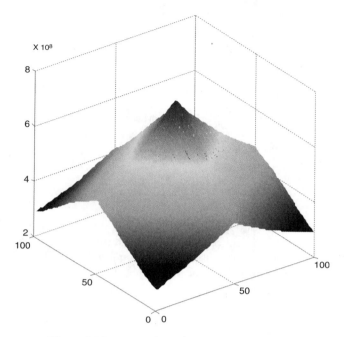

Figure 3.24   2D autocorrelation function of image

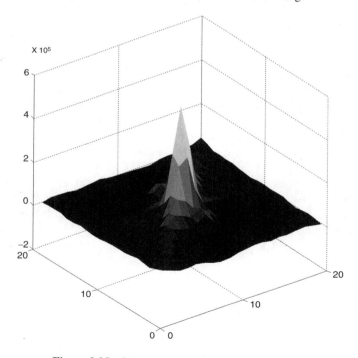

Figure 3.25   2D autocorrelation function of residual

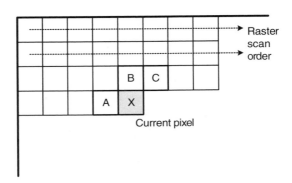

Figure 3.26   Spatial prediction (DPCM)

model is to decorrelate image or residual data further and to convert it into a form that can be efficiently compressed using an entropy coder. Practical image models typically have three main components, transformation (decorrelates and compacts the data), quantisation (reduces the precision of the transformed data) and reordering (arranges the data to group together significant values).

3.4.1 Predictive Image Coding

Motion compensation is an example of predictive coding in which an encoder creates a prediction of a region of the current frame based on a previous (or future) frame and subtracts this prediction from the current region to form a residual. If the prediction is successful, the energy in the residual is lower than in the original frame and the residual can be represented with fewer bits.

In a similar way, a prediction of an image sample or region may be formed from previously-transmitted samples in the same image or frame. Predictive coding was used as the basis for early image compression algorithms and is an important component of H.264 Intra coding (applied in the transform domain, see Chapter 6). Spatial prediction is sometimes described as 'Differential Pulse Code Modulation' (DPCM), a term borrowed from a method of differentially encoding PCM samples in telecommunication systems.

Figure 3.26 shows a pixel X that is to be encoded. If the frame is processed in raster order, then pixels A, B and C (neighbouring pixels in the current and previous rows) are available in both the encoder and the decoder (since these should already have been decoded before X). The encoder forms a prediction for X based on some combination of previously-coded pixels, subtracts this prediction from X and encodes the residual (the result of the subtraction). The decoder forms the same prediction and adds the decoded residual to reconstruct the pixel.

> ### *Example*
> Encoder prediction $P(X) = (2A + B + C)/4$
> Residual $R(X) = X - P(X)$ is encoded and transmitted.
>
> Decoder decodes $R(X)$ and forms the same prediction: $P(X) = (2A + B + C)/4$
> Reconstructed pixel $X = R(X) + P(X)$

If the encoding process is lossy (e.g. if the residual is quantised – see section 3.4.3) then the decoded pixels A', B' and C' may not be identical to the original A, B and C (due to losses during encoding) and so the above process could lead to a cumulative mismatch (or 'drift') between the encoder and decoder. In this case, the encoder should itself decode the residual R'(X) and reconstruct each pixel.

The encoder uses decoded pixels A', B' and C' to form the prediction, i.e. $P(X) = (2A' + B' + C')/4$ in the above example. In this way, both encoder and decoder use the same prediction P(X) and drift is avoided.

The compression efficiency of this approach depends on the accuracy of the prediction P(X). If the prediction is accurate (P(X) is a close approximation of X) then the residual energy will be small. However, it is usually not possible to choose a predictor that works well for all areas of a complex image and better performance may be obtained by adapting the predictor depending on the local statistics of the image (for example, using different predictors for areas of flat texture, strong vertical texture, strong horizontal texture, etc.). It is necessary for the encoder to indicate the choice of predictor to the decoder and so there is a tradeoff between efficient prediction and the extra bits required to signal the choice of predictor.

3.4.2 Transform Coding

3.4.2.1 Overview

The purpose of the transform stage in an image or video CODEC is to convert image or motion-compensated residual data into another domain (the transform domain). The choice of transform depends on a number of criteria:

1. Data in the transform domain should be decorrelated (separated into components with minimal inter-dependence) and compact (most of the energy in the transformed data should be concentrated into a small number of values).
2. The transform should be reversible.
3. The transform should be computationally tractable (low memory requirement, achievable using limited-precision arithmetic, low number of arithmetic operations, etc.).

Many transforms have been proposed for image and video compression and the most popular transforms tend to fall into two categories: block-based and image-based. Examples of block-based transforms include the Karhunen–Loeve Transform (KLT), Singular Value Decomposition (SVD) and the ever-popular Discrete Cosine Transform (DCT) [3]. Each of these operate on blocks of $N \times N$ image or residual samples and hence the image is processed in units of a block. Block transforms have low memory requirements and are well-suited to compression of block-based motion compensation residuals but tend to suffer from artefacts at block edges ('blockiness'). Image-based transforms operate on an entire image or frame (or a large section of the image known as a 'tile'). The most popular image transform is the Discrete Wavelet Transform (DWT or just 'wavelet'). Image transforms such as the DWT have been shown to out-perform block transforms for still image compression but they tend to have higher memory requirements (because the whole image or tile is processed as a unit) and do not 'fit' well with block-based motion compensation. The DCT and the DWT both feature in MPEG-4 Visual (and a variant of the DCT is incorporated in H.264) and are discussed further in the following sections.

3.4.2.2 DCT

The Discrete Cosine Transform (DCT) operates on $\mathbf{X}$, a block of $N \times N$ samples (typically image samples or residual values after prediction) and creates $\mathbf{Y}$, an $N \times N$ block of coefficients. The action of the DCT (and its inverse, the IDCT) can be described in terms of a transform matrix $\mathbf{A}$. The forward DCT (FDCT) of an $N \times N$ sample block is given by:

$$\mathbf{Y} = \mathbf{AXA}^{\mathrm{T}} \tag{3.1}$$

and the inverse DCT (IDCT) by:

$$\mathbf{X} = \mathbf{A}^{\mathrm{T}}\mathbf{YA} \tag{3.2}$$

where $\mathbf{X}$ is a matrix of samples, $\mathbf{Y}$ is a matrix of coefficients and $\mathbf{A}$ is an $N \times N$ transform matrix. The elements of $\mathbf{A}$ are:

$$A_{ij} = C_i \cos \frac{(2j+1)i\pi}{2N} \quad \text{where } C_i = \sqrt{\frac{1}{N}} \ (i=0), \quad C_i = \sqrt{\frac{2}{N}} \ (i>0) \tag{3.3}$$

Equation 3.1 and equation 3.2 may be written in summation form:

$$Y_{xy} = C_x C_y \sum_{i=0}^{N-1} \sum_{j=0}^{N-1} X_{ij} \cos \frac{(2j+1)y\pi}{2N} \cos \frac{(2i+1)x\pi}{2N} \tag{3.4}$$

$$X_{ij} = \sum_{x=0}^{N-1} \sum_{y=0}^{N-1} C_x C_y Y_{xy} \cos \frac{(2j+1)y\pi}{2N} \cos \frac{(2i+1)x\pi}{2N} \tag{3.5}$$

Example: $N = 4$

The transform matrix $\mathbf{A}$ for a 4×4 DCT is:

$$\mathbf{A} = \begin{bmatrix} \frac{1}{2}\cos(0) & \frac{1}{2}\cos(0) & \frac{1}{2}\cos(0) & \frac{1}{2}\cos(0) \\ \sqrt{\frac{1}{2}}\cos\left(\frac{\pi}{8}\right) & \sqrt{\frac{1}{2}}\cos\left(\frac{3\pi}{8}\right) & \sqrt{\frac{1}{2}}\cos\left(\frac{5\pi}{8}\right) & \sqrt{\frac{1}{2}}\cos\left(\frac{7\pi}{8}\right) \\ \sqrt{\frac{1}{2}}\cos\left(\frac{2\pi}{8}\right) & \sqrt{\frac{1}{2}}\cos\left(\frac{6\pi}{8}\right) & \sqrt{\frac{1}{2}}\cos\left(\frac{10\pi}{8}\right) & \sqrt{\frac{1}{2}}\cos\left(\frac{14\pi}{8}\right) \\ \sqrt{\frac{1}{2}}\cos\left(\frac{3\pi}{8}\right) & \sqrt{\frac{1}{2}}\cos\left(\frac{9\pi}{8}\right) & \sqrt{\frac{1}{2}}\cos\left(\frac{15\pi}{8}\right) & \sqrt{\frac{1}{2}}\cos\left(\frac{21\pi}{8}\right) \end{bmatrix} \tag{3.6}$$

The cosine function is symmetrical and repeats after 2π radians and hence **A** can be simplified to:

$$
\mathbf{A} = \begin{bmatrix}
\dfrac{1}{2} & \dfrac{1}{2} & \dfrac{1}{2} & \dfrac{1}{2} \\[2mm]
\sqrt{\dfrac{1}{2}}\cos\left(\dfrac{\pi}{8}\right) & \sqrt{\dfrac{1}{2}}\cos\left(\dfrac{3\pi}{8}\right) & -\sqrt{\dfrac{1}{2}}\cos\left(\dfrac{3\pi}{8}\right) & -\sqrt{\dfrac{1}{2}}\cos\left(\dfrac{\pi}{8}\right) \\[2mm]
\dfrac{1}{2} & -\dfrac{1}{2} & -\dfrac{1}{2} & \dfrac{1}{2} \\[2mm]
\sqrt{\dfrac{1}{2}}\cos\left(\dfrac{3\pi}{8}\right) & -\sqrt{\dfrac{1}{2}}\cos\left(\dfrac{\pi}{8}\right) & \sqrt{\dfrac{1}{2}}\cos\left(\dfrac{\pi}{8}\right) & -\sqrt{\dfrac{1}{2}}\cos\left(\dfrac{3\pi}{8}\right)
\end{bmatrix}
\tag{3.7}
$$

or

$$
\mathbf{A} = \begin{bmatrix}
a & a & a & a \\
b & c & -c & -b \\
a & -a & -a & a \\
c & -b & b & c
\end{bmatrix}
\quad \text{where} \quad
\begin{aligned}
a &= \frac{1}{2} \\
b &= \sqrt{\frac{1}{2}}\cos\left(\frac{\pi}{8}\right) \\
c &= \sqrt{\frac{1}{2}}\cos\left(\frac{3\pi}{8}\right)
\end{aligned}
\tag{3.8}
$$

Evaluating the cosines gives:

$$
\mathbf{A} = \begin{bmatrix}
0.5 & 0.5 & 0.5 & 0.5 \\
0.653 & 0.271 & 0.271 & -0.653 \\
0.5 & -0.5 & -0.5 & 0.5 \\
0.271 & -0.653 & -0.653 & 0.271
\end{bmatrix}
$$

The output of a two-dimensional FDCT is a set of $N \times N$ coefficients representing the image block data in the DCT domain and these coefficients can be considered as 'weights' of a set of standard *basis patterns*. The basis patterns for the 4×4 and 8×8 DCTs are shown in Figure 3.27 and Figure 3.28 respectively and are composed of combinations of horizontal and vertical cosine functions. Any image block may be reconstructed by combining all $N \times N$ basis patterns, with each basis multiplied by the appropriate weighting factor (coefficient).

Example 1 Calculating the DCT of a 4×4 block

X is 4×4 block of samples from an image:

	$j = 0$	1	2	3
$i = 0$	5	11	8	10
1	9	8	4	12
2	1	10	11	4
3	19	6	15	7

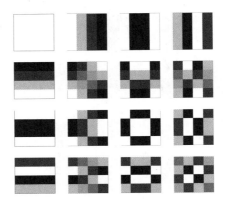

Figure 3.27   4 × 4 DCT basis patterns

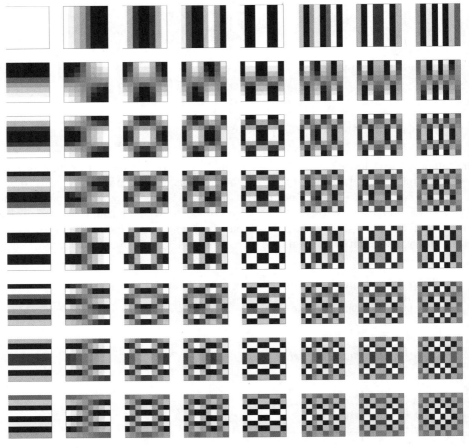

Figure 3.28   8 × 8 DCT basis patterns

The Forward DCT of $\mathbf{X}$ is given by: $\mathbf{Y} = \mathbf{AXA}^T$. The first matrix multiplication, $\mathbf{Y}' = \mathbf{AX}$, corresponds to calculating the one-dimensional DCT of each *column* of $\mathbf{X}$. For example, Y'_{00} is calculated as follows:

$$Y'_{00} = A_{00}X_{00} + A_{01}X_{10} + A_{02}X_{20} + A_{03}X_{30} = (0.5 * 5) + (0.5 * 9) + (0.5 * 1)$$
$$+ (0.5 * 19) = 17.0$$

The complete result of the column calculations is:

$$\mathbf{Y}' = \mathbf{AX} = \begin{bmatrix} 17 & 17.5 & 19 & 16.5 \\ -6.981 & 2.725 & -6.467 & 4.125 \\ 7 & -0.5 & 4 & 0.5 \\ -9.015 & 2.660 & 2.679 & -4.414 \end{bmatrix}$$

Carrying out the second matrix multiplication, $\mathbf{Y} = \mathbf{Y}'\mathbf{A}^T$, is equivalent to carrying out a 1-D DCT on each *row* of $\mathbf{Y}'$:

$$\mathbf{Y} = \mathbf{AXA}^T = \begin{bmatrix} 35.0 & -0.079 & -1.5 & 1.115 \\ -3.299 & -4.768 & 0.443 & -9.010 \\ 5.5 & 3.029 & 2.0 & 4.699 \\ -4.045 & -3.010 & -9.384 & -1.232 \end{bmatrix}$$

(Note: the order of the row and column calculations does not affect the final result).

Example 2 Image block and DCT coefficients

Figure 3.29 shows an image with a 4×4 block selected and Figure 3.30 shows the block in close-up, together with the DCT coefficients. The advantage of representing the block in the DCT domain is not immediately obvious since there is no reduction in the amount of data; instead of 16 pixel values, we need to store 16 DCT coefficients. The usefulness of the DCT becomes clear when the block is reconstructed from a subset of the coefficients.

Figure 3.29 Image section showing 4×4 block

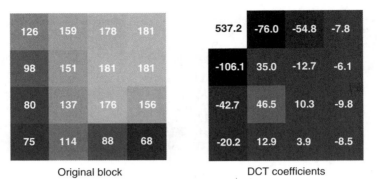

Original block   DCT coefficients

Figure 3.30   Close-up of 4 × 4 block; DCT coefficients

Setting all the coefficients to zero except the most significant (coefficient 0,0, described as the 'DC' coefficient) and performing the IDCT gives the output block shown in Figure 3.31(a), the mean of the original pixel values. Calculating the IDCT of the two most significant coefficients gives the block shown in Figure 3.31(b). Adding more coefficients before calculating the IDCT produces a progressively more accurate reconstruction of the original block and by the time five coefficients are included (Figure 3.31(d)), the reconstructed block is a reasonably close match to the original. Hence it is possible to reconstruct an approximate copy of the block from a subset of the 16 DCT coefficients. Removing the coefficients with insignificant magnitudes (for example by quantisation, see Section 3.4.3) enables image data to be represented with a reduced number of coefficient values at the expense of some loss of quality.

3.4.2.3 Wavelet

The popular 'wavelet transform' (widely used in image compression is based on sets of filters with coefficients that are equivalent to discrete wavelet functions [4]. The basic operation of a discrete wavelet transform is as follows, applied to a discrete signal containing N samples. A pair of filters are applied to the signal to decompose it into a low frequency band (L) and a high frequency band (H). Each band is subsampled by a factor of two, so that the two frequency bands each contain $N/2$ samples. With the correct choice of filters, this operation is reversible.

This approach may be extended to apply to a two-dimensional signal such as an intensity image (Figure 3.32). Each row of a 2D image is filtered with a low-pass and a high-pass filter (L_x and H_x) and the output of each filter is down-sampled by a factor of two to produce the intermediate images L and H. L is the original image low-pass filtered and downsampled in the x-direction and H is the original image high-pass filtered and downsampled in the x-direction. Next, each column of these new images is filtered with low- and high-pass filters (L_y and H_y) and down-sampled by a factor of two to produce four sub-images (LL, LH, HL and HH). These four 'sub-band' images can be combined to create an output image with the same number of samples as the original (Figure 3.33). 'LL' is the original image, low-pass filtered in horizontal and vertical directions and subsampled by a factor of 2. 'HL' is high-pass filtered in the vertical direction and contains residual vertical frequencies, 'LH' is high-pass filtered in the horizontal direction and contains residual horizontal frequencies and 'HH' is high-pass filtered in both horizontal and vertical directions. Between them, the four subband

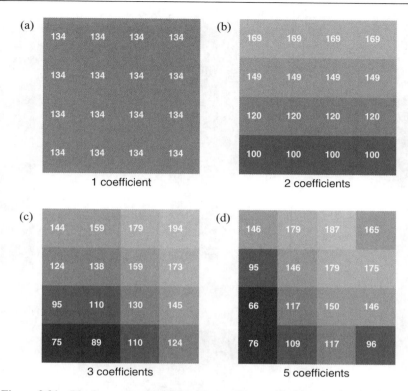

Figure 3.31   Block reconstructed from (a) one, (b) two, (c) three, (d) five coefficients

images contain all of the information present in the original image but the sparse nature of the LH, HL and HH subbands makes them amenable to compression.

In an image compression application, the two-dimensional wavelet decomposition described above is applied again to the 'LL' image, forming four new subband images. The resulting low-pass image (always the top-left subband image) is iteratively filtered to create a tree of subband images. Figure 3.34 shows the result of two stages of this decomposition and Figure 3.35 shows the result of five stages of decomposition. Many of the samples (coefficients) in the higher-frequency subband images are close to zero (near-black) and it is possible to achieve compression by removing these insignificant coefficients prior to transmission. At the decoder, the original image is reconstructed by repeated up-sampling, filtering and addition (reversing the order of operations shown in Figure 3.32).

3.4.3 Quantisation

A quantiser maps a signal with a range of values X to a quantised signal with a reduced range of values Y. It should be possible to represent the quantised signal with fewer bits than the original since the range of possible values is smaller. A *scalar quantiser* maps one sample of the input signal to one quantised output value and a *vector quantiser* maps a group of input samples (a 'vector') to a group of quantised values.

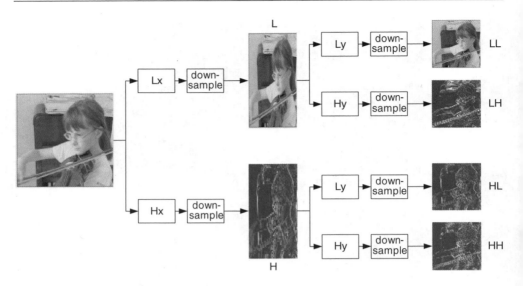

Figure 3.32   Two-dimensional wavelet decomposition process

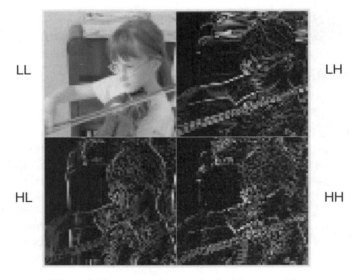

Figure 3.33   Image after one level of decomposition

3.4.3.1 Scalar Quantisation

A simple example of scalar quantisation is the process of rounding a fractional number to the nearest integer, i.e. the mapping is from R to Z. The process is lossy (not reversible) since it is not possible to determine the exact value of the original fractional number from the rounded integer.

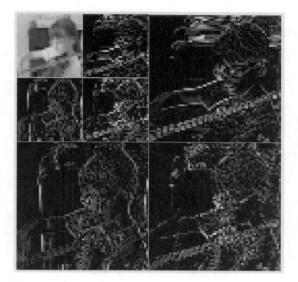

Figure 3.34 Two-stage wavelet decomposition of image

Figure 3.35 Five-stage wavelet decomposition of image

A more general example of a uniform quantiser is:

$$FQ = round\left(\frac{X}{QP}\right)$$
$$Y = FQ.QP$$

(3.9)

where QP is a quantisation 'step size'. The quantised output levels are spaced at uniform intervals of QP (as shown in the following example).

Example $Y = QP.round(X/QP)$

X			Y	
	$QP = 1$	$QP = 2$	$QP = 3$	$QP = 5$
-4	-4	-4	-3	-5
-3	-3	-2	-3	-5
-2	-2	-2	-3	0
-1	-1	0	0	0
0	0	0	0	0
1	1	0	0	0
2	2	2	3	0
3	3	2	3	5
4	4	4	3	5
5	5	4	6	5
6	6	6	6	5
7	7	6	6	5
8	8	8	9	10
9	9	8	9	10
10	10	10	9	10
11	11	10	12	10
......				

Figure 3.36 shows two examples of scalar quantisers, a linear quantiser (with a linear mapping between input and output values) and a nonlinear quantiser that has a 'dead zone' about zero (in which small-valued inputs are mapped to zero).

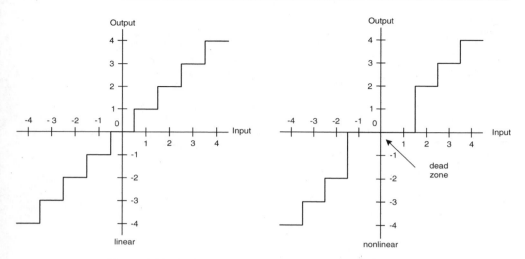

Figure 3.36 Scalar quantisers: linear; nonlinear with dead zone

In image and video compression CODECs, the quantisation operation is usually made up of two parts: a forward quantiser FQ in the encoder and an 'inverse quantiser' or (IQ) in the decoder (in fact quantization is not reversible and so a more accurate term is 'scaler' or 'rescaler'). A critical parameter is the *step size QP* between successive re-scaled values. If the step size is large, the range of quantised values is small and can therefore be efficiently represented (highly compressed) during transmission, but the re-scaled values are a crude approximation to the original signal. If the step size is small, the re-scaled values match the original signal more closely but the larger range of quantised values reduces compression efficiency.

Quantisation may be used to reduce the precision of image data after applying a transform such as the DCT or wavelet transform removing remove insignificant values such as near-zero DCT or wavelet coefficients. The forward quantiser in an image or video encoder is designed to map insignificant coefficient values to zero whilst retaining a reduced number of significant, nonzero coefficients. The output of a forward quantiser is typically a 'sparse' array of quantised coefficients, mainly containing zeros.

3.4.3.2 Vector Quantisation

A vector quantiser maps a set of input data (such as a block of image samples) to a single value (codeword) and, at the decoder, each codeword maps to an approximation to the original set of input data (a 'vector'). The set of vectors are stored at the encoder and decoder in a codebook. A typical application of vector quantisation to image compression [5] is as follows:

1. Partition the original image into regions (e.g. $M \times N$ pixel blocks).
2. Choose a vector from the codebook that matches the current region as closely as possible.
3. Transmit an index that identifies the chosen vector to the decoder.
4. At the decoder, reconstruct an approximate copy of the region using the selected vector.

A basic system is illustrated in Figure 3.37. Here, quantisation is applied in the spatial domain (i.e. groups of image samples are quantised as vectors) but it could equally be applied to

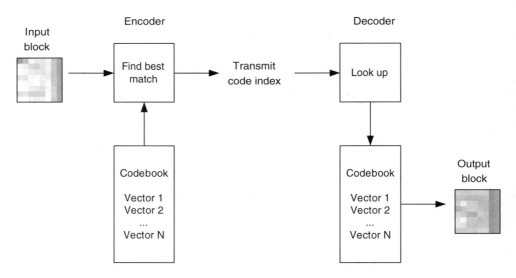

Figure 3.37   Vector quantisation

motion compensated and/or transformed data. Key issues in vector quantiser design include the design of the codebook and efficient searching of the codebook to find the optimal vector.

3.4.4 Reordering and Zero Encoding

Quantised transform coefficients are required to be encoded as compactly as possible prior to storage and transmission. In a transform-based image or video encoder, the output of the quantiser is a sparse array containing a few nonzero coefficients and a large number of zero-valued coefficients. Reordering (to group together nonzero coefficients) and efficient representation of zero coefficients are applied prior to entropy encoding. These processes are described for the DCT and wavelet transform.

3.4.4.1 DCT

Coefficient Distribution
The significant DCT coefficients of a block of image or residual samples are typically the 'low frequency' positions around the DC (0,0) coefficient. Figure 3.38 plots the probability of nonzero DCT coefficients at each position in an 8 × 8 block in a QCIF residual *frame* (Figure 3.6). The nonzero DCT coefficients are clustered around the top-left (DC) coefficient and the distribution is roughly symmetrical in the horizontal and vertical directions. For a residual *field* (Figure 3.39), Figure 3.40 plots the probability of nonzero DCT coefficients; here, the coefficients are clustered around the DC position but are 'skewed', i.e. more nonzero

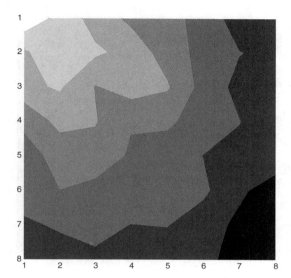

Figure 3.38 8×8 DCT coefficient distribution (frame)

Figure 3.39   Residual field picture

coefficients occur along the left-hand edge of the plot. This is because the field picture has a stronger high-frequency component in the vertical axis (due to the subsampling in the vertical direction) resulting in larger DCT coefficients corresponding to vertical frequencies (refer to Figure 3.27).

Scan

After quantisation, the DCT coefficients for a block are reordered to group together nonzero coefficients, enabling efficient representation of the remaining zero-valued quantised coefficients. The optimum reordering path (scan order) depends on the distribution of nonzero DCT coefficients. For a typical frame block with a distribution similar to Figure 3.38, a suitable scan order is a zigzag starting from the DC (top-left) coefficient. Starting with the DC coefficient, each quantised coefficient is copied into a one-dimensional array in the order shown in Figure 3.41. Nonzero coefficients tend to be grouped together at the start of the reordered array, followed by long sequences of zeros.

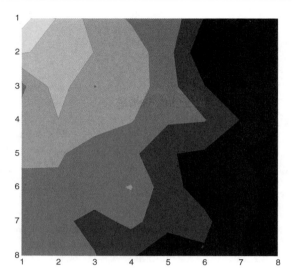

Figure 3.40   8 × 8 DCT coefficient distribution (field)

The zig-zag scan may not be ideal for a field block because of the skewed coefficient distribution (Figure 3.40) and a modified scan order such as Figure 3.42 may be more effective, in which coefficients on the left-hand side of the block are scanned before those on the right-hand side.

Run-Level Encoding

The output of the reordering process is an array that typically contains one or more clusters of nonzero coefficients near the start, followed by strings of zero coefficients. The large number of zero values may be encoded to represent them more compactly, for example by representing the array as a series of (run, level) pairs where *run* indicates the number of zeros preceding a nonzero coefficient and *level* indicates the magnitude of the nonzero coefficient.

Example

Input array: 16,0,0,−3,5,6,0,0,0,0,−7,...
Output values: (0,16),(2,−3),(0,5),(0,6),(4,−7)...

Each of these output values (a run-level pair) is encoded as a separate symbol by the entropy encoder.

Higher-frequency DCT coefficients are very often quantised to zero and so a reordered block will usually end in a run of zeros. A special case is required to indicate the final nonzero coefficient in a block. In so-called 'Two-dimensional' run-level encoding is used, each run-level pair is encoded as above and a separate code symbol, '*last*', indicates the end of the nonzero values. If 'Three-dimensional' run-level encoding is used, each symbol encodes

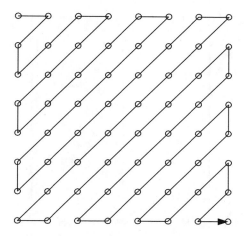

Figure 3.41 Zigzag scan order (frame block)

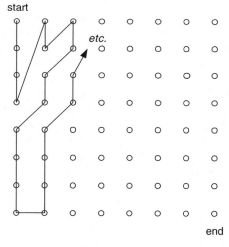

Figure 3.42 Zigzag scan order (field block)

three quantities, *run*, *level* and *last*. In the example above, if –7 is the final nonzero coefficient, the 3D values are:

$$(0, 16, 0), (2, -3, 0), (0, 5, 0), (0, 6, 0), (4, -7, 1)$$

The 1 in the final code indicates that this is the last nonzero coefficient in the block.

3.4.4.2 *Wavelet*

Coefficient Distribution

Figure 3.35 shows a typical distribution of 2D wavelet coefficients. Many coefficients in higher sub-bands (towards the bottom-right of the figure) are near zero and may be quantised

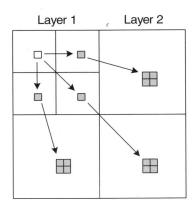

Figure 3.43   Wavelet coefficient and 'children'

to zero without significant loss of image quality. Nonzero coefficients tend to correspond to structures in the image; for example, the violin bow appears as a clear horizontal structure in all the horizontal and diagonal subbands. When a coefficient in a lower-frequency subband is nonzero, there is a strong probability that coefficients in the corresponding position in higher-frequency subbands will also be nonzero. We may consider a 'tree' of nonzero quantised coefficients, starting with a 'root' in a low-frequency subband. Figure 3.43 illustrates this concept. A single coefficient in the LL band of layer 1 has one corresponding coefficient in each of the other bands of layer 1 (i.e. these four coefficients correspond to the same region in the original image). The layer 1 coefficient position maps to four corresponding child coefficient positions in each subband at layer 2 (recall that the layer 2 subbands have twice the horizontal and vertical resolution of the layer 1 subbands).

Zerotree Encoding

It is desirable to encode the nonzero wavelet coefficients as compactly as possible prior to entropy coding [6]. An efficient way of achieving this is to encode each tree of nonzero coefficients starting from the lowest (root) level of the decomposition. A coefficient at the lowest layer is encoded, followed by its child coefficients at the next higher layer, and so on. The encoding process continues until the tree reaches a zero-valued coefficient. Further children of a zero valued coefficient are likely to be zero themselves and so the remaining children are represented by a single code that identifies a tree of zeros (*zerotree*). The decoder reconstructs the coefficient map starting from the root of each tree; nonzero coefficients are decoded and reconstructed and when a zerotree code is reached, all remaining 'children' are set to zero. This is the basis of the *embedded zero tree* (EZW) method of encoding wavelet coefficients. An extra possibility is included in the encoding process, where a zero coefficient may be followed by (a) a zero tree (as before) or (b) a nonzero child coefficient. Case (b) does not occur very often but reconstructed image quality is slightly improved by catering for the occasional occurrences of case (b).

3.5 ENTROPY CODER

The entropy encoder converts a series of symbols representing elements of the video sequence into a compressed bitstream suitable for transmission or storage. Input symbols may include quantised transform coefficients (run-level or zerotree encoded as described in Section 3.4.4), motion vectors (an x and y displacement vector for each motion-compensated block, with integer or sub-pixel resolution), markers (codes that indicate a resynchronisation point in the sequence), headers (macroblock headers, picture headers, sequence headers, etc.) and supplementary information ('side' information that is not essential for correct decoding). In this section we discuss methods of predictive pre-coding (to exploit correlation in local regions of the coded frame) followed by two widely-used entropy coding techniques, 'modified Huffman' variable length codes and arithmetic coding.

3.5.1 Predictive Coding

Certain symbols are highly correlated in local regions of the picture. For example, the average or DC value of neighbouring intra-coded blocks of pixels may be very similar; neighbouring motion vectors may have similar x and y displacements and so on. Coding efficiency may be improved by predicting elements of the current block or macroblock from previously-encoded data and encoding the difference between the prediction and the actual value.

The motion vector for a block or macroblock indicates the offset to a prediction reference in a previously-encoded frame. Vectors for neighbouring blocks or macroblocks are often correlated because object motion may extend across large regions of a frame. This is especially true for small block sizes (e.g. 4×4 block vectors, see Figure 3.22) and/or large moving objects. Compression of the motion vector field may be improved by predicting each motion vector from previously-encoded vectors. A simple prediction for the vector of the current macroblock X is the horizontally adjacent macroblock A (Figure 3.44), alternatively three or more previously-coded vectors may be used to predict the vector at macroblock X (e.g. A, B and C in Figure 3.44). The difference between the predicted and actual motion vector (Motion Vector Difference or MVD) is encoded and transmitted.

The quantisation parameter or quantiser step size controls the tradeoff between compression efficiency and image quality. In a real-time video CODEC it may be necessary to modify the quantisation within an encoded frame (for example to alter the compression ratio in order to match the coded bit rate to a transmission channel rate). It is usually

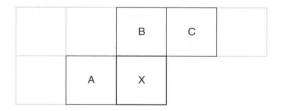

Figure 3.44 Motion vector prediction candidates

sufficient (and desirable) to change the parameter only by a small amount between successive coded macroblocks. The modified quantisation parameter must be signalled to the decoder and instead of sending a new quantisation parameter value, it may be preferable to send a delta or difference value (e.g. ± 1 or ± 2) indicating the change required. Fewer bits are required to encode a small delta value than to encode a completely new quantisation parameter.

3.5.2 Variable-length Coding

A variable-length encoder maps input symbols to a series of codewords (variable length codes or VLCs). Each symbol maps to a codeword and codewords may have varying length but must each contain an integral number of bits. Frequently-occurring symbols are represented with short VLCs whilst less common symbols are represented with long VLCs. Over a sufficiently large number of encoded symbols this leads to compression of the data.

3.5.2.1 Huffman Coding

Huffman coding assigns a VLC to each symbol based on the probability of occurrence of different symbols. According to the original scheme proposed by Huffman in 1952 [7], it is necessary to calculate the probability of occurrence of each symbol and to construct a set of variable length codewords. This process will be illustrated by two examples.

Example 1: Huffman coding, sequence 1 motion vectors

The motion vector difference data (MVD) for a video sequence ('sequence 1') is required to be encoded. Table 3.2 lists the probabilities of the most commonly-occurring motion vectors in the encoded sequence and their *information content*, $\log_2(1/p)$. To achieve optimum compression, each value should be represented with exactly $\log_2(1/p)$ bits. '0' is the most common value and the probability drops for larger motion vectors (this distribution is representative of a sequence containing moderate motion).

Table 3.2 Probability of occurrence of motion vectors in sequence 1

Vector	Probability p	$\log2(1/p)$
-2	0.1	3.32
-1	0.2	2.32
0	0.4	1.32
1	0.2	2.32
2	0.1	3.32

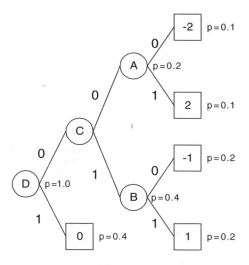

Figure 3.45 Generating the Huffman code tree: sequence 1 motion vectors

1. Generating the Huffman Code Tree

To generate a Huffman code table for this set of data, the following iterative procedure is carried out:

1. Order the list of data in increasing order of probability.
2. Combine the two lowest-probability data items into a 'node' and assign the joint probability of the data items to this node.
3. Re-order the remaining data items and node(s) in increasing order of probability and repeat step 2.

The procedure is repeated until there is a single 'root' node that contains all other nodes and data items listed 'beneath' it. This procedure is illustrated in Figure 3.45.

Original list:	The data items are shown as square boxes. Vectors (−2) and (+2) have the lowest probability and these are the first candidates for merging to form node 'A'.
Stage 1:	The newly-created node 'A' (shown as a circle) has a probability of 0.2 (from the combined probabilities of (−2) and (2)). There are now three items with probability 0.2. Choose vectors (−1) and (1) and merge to form node 'B'.
Stage 2:	A now has the lowest probability (0.2) followed by B and the vector 0; choose A and B as the next candidates for merging (to form 'C').
Stage 3:	Node C and vector (0) are merged to form 'D'.
Final tree:	The data items have all been incorporated into a binary 'tree' containing five data values and four nodes. Each data item is a 'leaf' of the tree.

2. Encoding

Each 'leaf' of the binary tree is mapped to a variable-length code. To find this code, the tree is traversed from the root node (D in this case) to the leaf (data item). For every branch, a 0 or 1

Table 3.3   Huffman codes for sequence 1
motion vectors

Vector	Code	Bits (actual)	Bits (ideal)
0	1	1	1.32
1	011	3	2.32
−1	010	3	2.32
2	001	3	3.32
−2	000	3	3.32

Table 3.4   Probability of occurrence
of motion vectors in sequence 2

Vector	Probability	$\log_2(1/p)$
−2	0.02	5.64
−1	0.07	3.84
0	0.8	0.32
1	0.08	3.64
2	0.03	5.06

is appended to the code, 0 for an upper branch, 1 for a lower branch (shown in the final tree of Figure 3.45), giving the following set of codes (Table 3.3).

Encoding is achieved by transmitting the appropriate code for each data item. Note that once the tree has been generated, the codes may be stored in a look-up table.

High probability data items are assigned short codes (e.g. 1 bit for the most common vector '0'). However, the vectors (−2, 2, −1, 1) are each assigned three-bit codes (despite the fact that −1 and 1 have higher probabilities than −2 and 2). The lengths of the Huffman codes (each an integral number of bits) do not match the ideal lengths given by $\log_2(1/p)$. No code contains any other code as a prefix which means that, reading from the left-hand bit, each code is uniquely decodable.

For example, the series of vectors (1, 0, −2) would be transmitted as the binary sequence 0111000.

3. Decoding

In order to decode the data, the decoder must have a local copy of the Huffman code tree (or look-up table). This may be achieved by transmitting the look-up table itself or by sending the list of data and probabilities prior to sending the coded data. Each uniquely-decodeable code is converted back to the original data, for example:

011 is decoded as (1)

1 is decoded as (0)

000 is decoded as (−2).

Example 2: Huffman coding, sequence 2 motion vectors

Repeating the process described above for a second sequence with a different distribution of motion vector probabilities gives a different result. The probabilities are listed in Table 3.4 and note that the zero vector is much more likely to occur in this example (representative of a sequence with little movement).

Table 3.5 Huffman codes for sequence 2 motion vectors

Vector	Code	Bits (actual)	Bits (ideal)
0	1	1	0.32
1	01	2	3.64
−1	001	3	3.84
2	0001	4	5.06
−2	0000	4	5.64

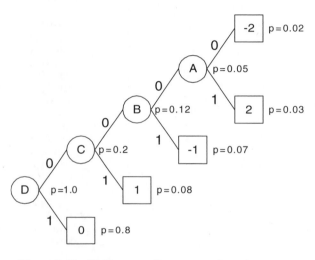

Figure 3.46 Huffman tree for sequence 2 motion vectors

The corresponding Huffman tree is given in Figure 3.46. The 'shape' of the tree has changed (because of the distribution of probabilities) and this gives a different set of Huffman codes (shown in Table 3.5). There are still four nodes in the tree, one less than the number of data items (five), as is always the case with Huffman coding.

If the probability distributions are accurate, Huffman coding provides a relatively compact representation of the original data. In these examples, the frequently occurring (0) vector is represented efficiently as a single bit. However, to achieve optimum compression, a separate code table is required for each of the two sequences because of their different probability distributions. The loss of potential compression efficiency due to the requirement for integral-length codes is very clear for vector '0' in sequence 2, since the optimum number of bits (information content) is 0.32 but the best that can be achieved with Huffman coding is 1 bit.

3.5.2.2 *Pre-calculated Huffman-based Coding*

The Huffman coding process has two disadvantages for a practical video CODEC. First, the decoder must use the same codeword set as the encoder. Transmitting the information contained in the probability table to the decoder would add extra overhead and reduce compression

Table 3.6 MPEG-4 Visual Transform Coefficient
(TCOEF) VLCs (partial, all codes <9 bits)

Last	Run	Level	Code
0	0	1	10s
0	1	1	110s
0	2	1	1110s
0	0	2	1111s
1	0	1	0111s
0	3	1	01101s
0	4	1	01100s
0	5	1	01011s
0	0	3	010101s
0	1	2	010100s
0	6	1	010011s
0	7	1	010010s
0	8	1	010001s
0	9	1	010000s
1	1	1	001111s
1	2	1	001110s
1	3	1	001101s
1	4	1	001100s
0	0	4	0010111s
0	10	1	0010110s
0	11	1	0010101s
0	12	1	0010100s
1	5	1	0010011s
1	6	1	0010010s
1	7	1	0010001s
1	8	1	0010000s
ESCAPE			0000011s
. . .	. . .	. . .	. . .

efficiency, particularly for shorter video sequences. Second, the probability table for a large video sequence (required to generate the Huffman tree) cannot be calculated until after the video data is encoded which may introduce an unacceptable delay into the encoding process. For these reasons, recent image and video coding standards define sets of codewords based on the probability distributions of 'generic' video material. The following two examples of pre-calculated VLC tables are taken from MPEG-4 Visual (Simple Profile).

Transform Coefficients (TCOEF)

MPEG-4 Visual uses 3D coding of quantised coefficients in which each codeword represents a combination of (run, level, last). A total of 102 specific combinations of (run, level, last) have VLCs assigned to them and 26 of these codes are shown in Table 3.6.

A further 76 VLCs are defined, each up to 13 bits long. The last bit of each codeword is the sign bit 's', indicating the sign of the decoded coefficient (0 = positive, 1 = negative). Any (run, level, last) combination that is not listed in the table is coded using an escape sequence, a special ESCAPE code (0000011) followed by a 13-bit fixed length code describing the values of run, level and last.

Table 3.7 MPEG4 Motion Vector
Difference (MVD) VLCs

MVD	Code
0	1
+0.5	010
−0.5	011
+1	0010
−1	0011
+1.5	00010
−1.5	00011
+2	0000110
−2	0000111
+2.5	00001010
−2.5	00001011
+3	00001000
−3	00001001
+3.5	00000110
−3.5	00000111
. . .	. . .

Some of the codes shown in Table 3.6 are represented in 'tree' form in Figure 3.47. A codeword containing a run of more than eight zeros is not valid, hence any codeword starting with 000000000. . . indicates an error in the bitstream (or possibly a start code, which begins with a long sequence of zeros, occurring at an unexpected position in the sequence). All other sequences of bits can be decoded as valid codes. Note that the smallest codes are allocated to short runs and small levels (e.g. code '10' represents a run of 0 and a level of ±1), since these occur most frequently.

Motion Vector Difference (MVD)

Differentially coded motion vectors (MVD) are each encoded as a pair of VLCs, one for the x-component and one for the y-component. Part of the table of VLCs is shown in Table 3.7. A further 49 codes (8–13 bits long) are not shown here. Note that the shortest codes represent small motion vector differences (e.g. MVD $= 0$ is represented by a single bit code '1').

These code tables are clearly similar to 'true' Huffman codes since each symbol is assigned a unique codeword, common symbols are assigned shorter codewords and, within a table, no codeword is the prefix of any other codeword. The main differences from 'true' Huffman coding are (1) the codewords are pre-calculated based on 'generic' probability distributions and (b) in the case of TCOEF, only 102 commonly-occurring symbols have defined codewords with any other symbol encoded using a fixed-length code.

3.5.2.3 *Other Variable-length Codes*

As well as Huffman and Huffman-based codes, a number of other families of VLCs are of potential interest in video coding applications. One serious disadvantage of Huffman-based codes for transmission of coded data is that they are sensitive to transmission errors. An

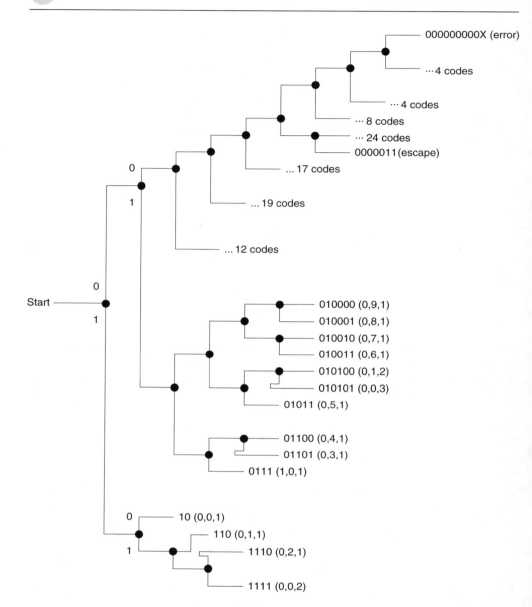

Figure 3.47   MPEG4 TCOEF VLCs (partial)

error in a sequence of VLCs may cause a decoder to lose synchronisation and fail to decode subsequent codes correctly, leading to spreading or propagation of an error in a decoded sequence. Reversible VLCs (RVLCs) that can be successfully decoded in either a forward or a backward direction can improve decoding performance when errors occur (see Section 5.3). A drawback of pre-defined code tables (such as Table 3.6 and Table 3.7) is that both encoder and decoder must store the table in some form. An alternative approach is to use codes that can be generated automatically ('on the fly') if the input symbol is known. Exponential Golomb codes (Exp-Golomb) fall into this category and are described in Chapter 6.

3.5.3 Arithmetic Coding

The variable length coding schemes described in Section 3.5.2 share the fundamental disadvantage that assigning a codeword containing an integral number of bits to each symbol is sub-optimal, since the optimal number of bits for a symbol depends on the information content and is usually a fractional number. Compression efficiency of variable length codes is particularly poor for symbols with probabilities greater than 0.5 as the best that can be achieved is to represent these symbols with a single-bit code.

Arithmetic coding provides a practical alternative to Huffman coding that can more closely approach theoretical maximum compression ratios [8]. An arithmetic encoder converts a sequence of data symbols into a single fractional number and can approach the optimal fractional number of bits required to represent each symbol.

Example

Table 3.8 lists the five motion vector values $(-2, -1, 0, 1, 2)$ and their probabilities from Example 1 in Section 3.5.2.1. Each vector is assigned a *sub-range* within the range 0.0 to 1.0, depending on its probability of occurrence. In this example, (-2) has a probability of 0.1 and is given the subrange 0–0.1 (i.e. the first 10% of the total range 0 to 1.0). (-1) has a probability of 0.2 and is given the next 20% of the total range, i.e. the subrange 0.1–0.3. After assigning a sub-range to each vector, the total range 0–1.0 has been divided amongst the data symbols (the vectors) according to their probabilities (Figure 3.48).

Table 3.8 Motion vectors, sequence 1:
probabilities and sub-ranges

Vector	Probability	$\log_2(1/P)$	Sub-range
−2	0.1	3.32	0–0.1
−1	0.2	2.32	0.1–0.3
0	0.4	1.32	0.3–0.7
1	0.2	2.32	0.7–0.9
2	0.1	3.32	0.9–1.0

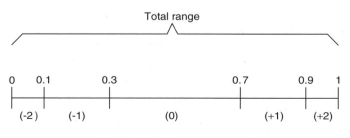

Figure 3.48 Sub-range example

Encoding Procedure for Vector Sequence (0, −1, 0, 2):

Encoding procedure	Range (L → H)	Symbol (L → H)	Sub-range	Notes
1. Set the initial range.	0 → 1.0			
2. For the first data symbol, find the corresponding sub-range (Low to High).		(0)	0.3 → 0.7	
3. Set the new range (1) to this sub-range.	0.3 → 0.7			
4. For the next data symbol, find the sub-range L to H.		(−1)	0.1 → 0.3	This is the sub-range within the interval 0–1
5. Set the new range (2) to this sub-range within the previous range.	0.34 → 0.42			0.34 is 10% of the range 0.42 is 30% of the range
6. Find the next sub-range.		(0)	0.3 → 0.7	
7. Set the new range (3) within the previous range.	0.364→0.396			0.364 is 30% of the range; 0.396 is 70% of the range
8. Find the next sub-range.		(2)	0.9 → 1.0	
9. Set the new range (4) within the previous range.	0.3928→0.396			0.3928 is 90% of the range; 0.396 is 100% of the range

Each time a symbol is encoded, the range (L to H) becomes progressively smaller. At the end of the encoding process (four steps in this example), we are left with a final range (L to H). The entire sequence of data symbols can be represented by transmitting any fractional number that lies within this final range. In the example above, we could send any number in the range 0.3928 to 0.396: for example, 0.394. Figure 3.49 shows how the initial range (0 to 1) is progressively partitioned into smaller ranges as each data symbol is processed. After encoding the first symbol (vector 0), the new range is (0.3, 0.7). The next symbol (vector −1) selects the sub-range (0.34, 0.42) which becomes the new range, and so on. The final symbol (vector +2) selects the sub-range (0.3928, 0.396) and the number 0.394 (falling within this range) is transmitted. 0.394 can be represented as a fixed-point fractional number using nine bits, so our data sequence (0, −1, 0, 2) is compressed to a nine-bit quantity.

Decoding Procedure

Decoding procedure	Range	Sub-range	Decoded symbol
1. Set the initial range.	0 → 1		
2. Find the sub-range in which the received number falls. This indicates the first data symbol.		0.3 → 0.7	(0)

(*cont.*)

Decoding procedure	Range	Sub-range	Decoded symbol
3. Set the new range (1) to this sub-range.	0.3 → 0.7		
4. Find the sub-range of *the new range* in which the received number falls. This indicates the second data symbol.		0.34 → 0.42	(−1)
5. Set the new range (2) to this sub-range within the previous range.	0.34 → 0.42		
6. Find the sub-range in which the received number falls and decode the third data symbol.		0.364 → 0.396	(0)
7. Set the new range (3) to this sub-range within the previous range.	0.364 → 0.396		
8. Find the sub-range in which the received number falls and decode the fourth data symbol.		0.3928→ 0.396	

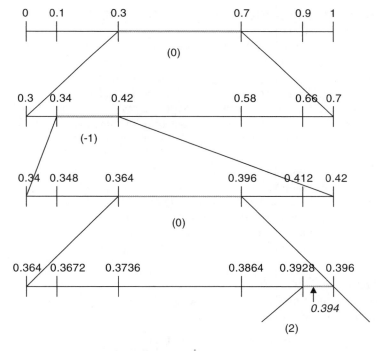

Figure 3.49 Arithmetic coding example

The principal advantage of arithmetic coding is that the transmitted number (0.394 in this case, which may be represented as a fixed-point number with sufficient accuracy using nine bits) is not constrained to an integral number of bits for each transmitted data symbol. To achieve optimal compression, the sequence of data symbols should be represented with:

$$\log_2(1/P_0) + \log_2(1/P_{-1}) + \log_2(1/P_0) + \log_2(1/P_2)\text{bits} = 8.28\text{bits}$$

In this example, arithmetic coding achieves nine bits, which is close to optimum. A scheme using an integral number of bits for each data symbol (such as Huffman coding) is unlikely to come so close to the optimum number of bits and, in general, arithmetic coding can out-perform Huffman coding.

3.5.3.1 Context-based Arithmetic Coding

Successful entropy coding depends on accurate models of symbol probability. Context-based Arithmetic Encoding (CAE) uses local spatial and/or temporal characteristics to estimate the probability of a symbol to be encoded. CAE is used in the JBIG standard for bi-level image compression [9] and has been adopted for coding binary shape 'masks' in MPEG-4 Visual (see Chapter 5) and entropy coding in the Main Profile of H.264 (see Chapter 6).

3.6 THE HYBRID DPCM/DCT VIDEO CODEC MODEL

The major video coding standards released since the early 1990s have been based on the same generic design (or model) of a video CODEC that incorporates a motion estimation and compensation front end (sometimes described as DPCM), a transform stage and an entropy encoder. The model is often described as a hybrid DPCM/DCT CODEC. Any CODEC that is compatible with H.261, H.263, MPEG-1, MPEG-2, MPEG-4 Visual and H.264 has to implement a similar set of basic coding and decoding functions (although there are many differences of detail between the standards and between implementations).

Figure 3.50 and Figure 3.51 show a generic DPCM/DCT hybrid encoder and decoder. In the encoder, video frame n (F_n) is processed to produce a coded (compressed) bitstream and in the decoder, the compressed bitstream (shown at the right of the figure) is decoded to produce a reconstructed video frame F'_n. not usually identical to the source frame. The figures have been deliberately drawn to highlight the common elements within encoder and decoder. Most of the functions of the decoder are actually contained within the encoder (the reason for this will be explained below).

Encoder Data Flow
There are two main data flow paths in the encoder, left to right (encoding) and right to left (reconstruction). The encoding flow is as follows:

1. An input video frame F_n is presented for encoding and is processed in units of a macroblock (corresponding to a 16×16 luma region and associated chroma samples).
2. F_n is compared with a *reference* frame, for example the previous encoded frame (F'_{n-1}). A motion estimation function finds a 16×16 region in F'_{n-1} (or a sub-sample interpolated

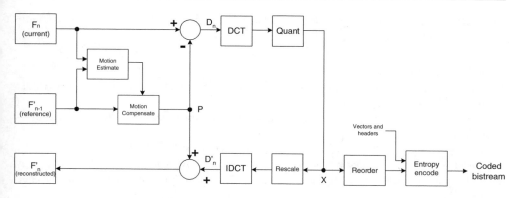

Figure 3.50   DPCM/DCT video encoder

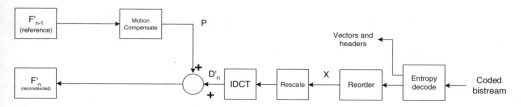

Figure 3.51   DPCM/DCT video decoder

version of F'$_{n-1}$) that 'matches' the current macroblock in F$_n$ (i.e. is similar according to some matching criteria). The offset between the current macroblock position and the chosen reference region is a motion vector MV.

3. Based on the chosen motion vector MV, a motion compensated prediction P is generated (the 16 × 16 region selected by the motion estimator).
4. P is subtracted from the current macroblock to produce a residual or difference macro-block D.
5. D is transformed using the DCT. Typically, D is split into 8 × 8 or 4 × 4 sub-blocks and each sub-block is transformed separately.
6. Each sub-block is quantised (X).
7. The DCT coefficients of each sub-block are reordered and run-level coded.
8. Finally, the coefficients, motion vector and associated header information for each mac-roblock are entropy encoded to produce the compressed bitstream.

The reconstruction data flow is as follows:

1. Each quantised macroblock X is rescaled and inverse transformed to produce a decoded residual D'. Note that the nonreversible quantisation process means that D' is not identical to D (i.e. distortion has been introduced).
2. The motion compensated prediction P is added to the residual D' to produce a reconstructed macroblock and the reconstructed macroblocks are saved to produce reconstructed frame F'$_n$.

After encoding a complete frame, the reconstructed frame F'_n may be used as a reference frame for the next encoded frame F_{n+1} .

Decoder Data Flow
1. A compressed bitstream is entropy decoded to extract coefficients, motion vector and header for each macroblock.
2. Run-level coding and reordering are reversed to produce a quantised, transformed macroblock X.
3. X is rescaled and inverse transformed to produce a decoded residual D'.
4. The decoded motion vector is used to locate a 16×16 region in the decoder's copy of the previous (reference) frame F'_{n-1}. This region becomes the motion compensated prediction P.
5. P is added to D' to produce a reconstructed macroblock. The reconstructed macroblocks are saved to produce decoded frame F'_n.

After a complete frame is decoded, F'_n is ready to be displayed and may also be stored as a reference frame for the next decoded frame F'_{n+1}.

It is clear from the figures and from the above explanation that the encoder includes a decoding path (rescale, IDCT, reconstruct). This is necessary to ensure that the encoder and decoder use identical reference frames F'_{n-1} for motion compensated prediction.

Example

A 25-Hz video sequence in CIF format (352×288 luminance samples and 176×144 red/blue chrominance samples per frame) is encoded and decoded using a DPCM/DCT CODEC. Figure 3.52 shows a CIF (video frame (F_n) that is to be encoded and Figure 3.53 shows the reconstructed previous frame F'_{n-1}. Note that F'_{n-1} has been encoded and decoded and shows some distortion. The difference between F_n and F'_{n-1} *without* motion compensation (Figure 3.54) clearly still contains significant energy, especially around the edges of moving areas.

Motion estimation is carried out with a 16×16 luma block size and half-sample accuracy, producing the set of vectors shown in Figure 3.55 (superimposed on the current frame for clarity). Many of the vectors are zero (shown as dots) which means that the best match for the current macroblock is in the same position in the reference frame. Around moving areas, the vectors tend to point in the direction *from which* blocks have moved (e.g. the man on the left is walking to the left; the vectors therefore point to the *right*, i.e. where he has come from). Some of the vectors do not appear to correspond to 'real' movement (e.g. on the surface of the table) but indicate simply that the best match is not at the same position in the reference frame. 'Noisy' vectors like these often occur in homogeneous regions of the picture, where there are no clear object features in the reference frame.

The motion-compensated reference frame (Figure 3.56) is the reference frame 'reorganized' according to the motion vectors. For example, note that the walking person (2nd left) has been moved to the left to provide a better match for the same person in the current frame and that the hand of the left-most person has been moved down to provide an improved match. Subtracting the motion compensated reference frame from the current frame gives the motion-compensated residual in Figure 3.57 in which the energy has clearly been reduced, particularly around moving areas.

Figure 3.52 Input frame F_n

Figure 3.53 Reconstructed reference frame F'_{n-1}

Figure 3.54   Residual $F_n - F'_{n-1}$ (no motion compensation)

Figure 3.55 16×16 motion vectors (superimposed on frame)

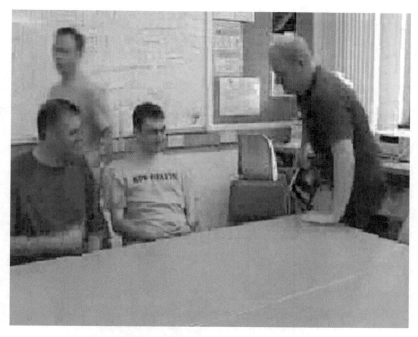

Figure 3.56 Motion compensated reference frame

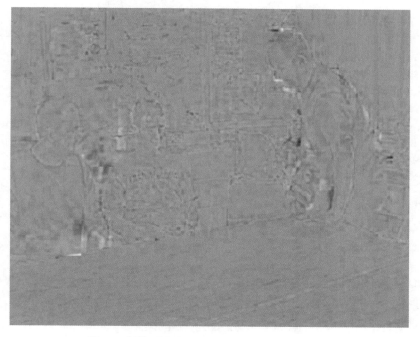

Figure 3.57 Motion compensated residual frame

Table 3.9   Residual luminance samples
(upper-right 8 × 8 block)

−4	−4	−1	0	1	1	0	−2
1	2	3	2	−1	−3	−6	−3
6	6	4	−4	−9	−5	−6	−5
10	8	−1	−4	−6	−1	2	4
7	9	−5	−9	−3	0	8	13
0	3	−9	−12	−8	−9	−4	1
−1	4	−9	−13	−8	−16	−18	−13
14	13	−1	−6	3	−5	−12	−7

Figure 3.58   Original macroblock (luminance)

Figure 3.58 shows a macroblock from the original frame (taken from around the head of the figure on the right) and Figure 3.59 the luminance residual after motion compensation. Applying a 2D DCT to the top-right 8 × 8 block of luminance samples (Table 3.9) produces the DCT coefficients listed in Table 3.10. The magnitude of each coefficient is plotted in Figure 3.60; note that the larger coefficients are clustered around the top-left (DC) coefficient.

A simple forward quantiser is applied:

$$Qcoeff = round(coeff/Qstep)$$

where $Qstep$ is the quantiser step size, 12 in this example. Small-valued coefficients become zero in the quantised block (Table 3.11) and the nonzero outputs are clustered around the top-left (DC) coefficient.

The quantised block is reordered in a zigzag scan (starting at the top-left) to produce a linear array:

$$-1, 2, 1, -1, -1, 2, 0, -1, 1, -1, 2, -1, -1, 0, 0, -1, 0, 0, 0, -1, -1, 0, 0, 0, 0, 0, 1, 0, \ldots$$

Table 3.10   DCT coefficients

−13.50	20.47	20.20	2.14	−0.50	−10.48	−3.50	−0.62
10.93	−11.58	−10.29	−5.17	−2.96	10.44	4.96	−1.26
−8.75	9.22	−17.19	2.26	3.83	−2.45	1.77	1.89
−7.10	−17.54	1.24	−0.91	0.47	−0.37	−3.55	0.88
19.00	−7.20	4.08	5.31	0.50	0.18	−0.61	0.40
−13.06	3.12	−2.04	−0.17	−1.19	1.57	−0.08	−0.51
1.73	−0.69	1.77	0.78	−1.86	1.47	1.19	0.42
−1.99	−0.05	1.24	−0.48	−1.86	−1.17	−0.21	0.92

Table 3.11   Quantized coefficients

−1	2	2	0	0	−1	0	0
1	−1	−1	0	0	1	0	0
−1	1	−1	0	0	0	0	0
−1	−1	0	0	0	0	0	0
2	−1	0	0	0	0	0	0
−1	0	0	0	0	0	0	0
0	0	0	0	0	0	0	0
0	0	0	0	0	0	0	0

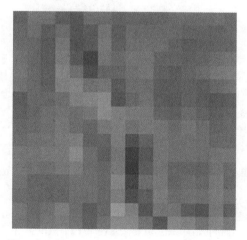

Figure 3.59   Residual macroblock (luminance)

This array is processed to produce a series of (zero run, level) pairs:

(0, −1)(0, 2)(0, 1)(0, −1)(0, −1)(0, 2)(1, −1)(0, 1)(0, −1)(0, 2)(0, −1)(0, −1)(2, −1)
(3, −1)(0, −1)(5, 1)(EOB)

'EOB' (End Of Block) indicates that the remainder of the coefficients are zero.

Each (run, level) pair is encoded as a VLC. Using the MPEG-4 Visual TCOEF table (Table 3.6), the VLCs shown in Table 3.12 are produced.

Table 3.12 Variable length coding example

Run, Level, Last	VLC (including sign bit)
(0, −1, 0)	101
(0, 2, 0)	11100
(0, 1, 0)	100
(0, −1, 0)	101
(0, −1, 0)	101
(0, 2, 0)	11100
(1, −1, 0)	1101
. . .	. . .
(5, 1, 1)	00100110

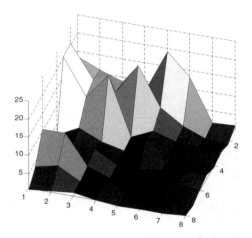

Figure 3.60 DCT coefficient magnitudes (top-right 8 × 8 block)

The final VLC signals that LAST = 1, indicating that this is the end of the block. The motion vector for this macroblock is (0, 1) (i.e. the vector points downwards). The predicted vector (based on neighbouring macroblocks) is (0,0) and so the motion vector difference values are MVDx = 0, MVDy = +1. Using the MPEG4 MVD table (Table 3.7), these are coded as (1) and (0010) respectively.

The macroblock is transmitted as a series of VLCs, including a macroblock header, motion vector difference (X and Y) and transform coefficients (TCOEF) for each 8 × 8 block.

At the decoder, the VLC sequence is decoded to extract header parameters, MVDx and MVDy and (run,level) pairs for each block. The 64-element array of reordered coefficients is reconstructed by inserting (run) zeros before every (level). The array is then ordered to produce an 8 × 8 block (identical to Table 3.11). The quantised coefficients are rescaled using:

$$Rcoeff = Qstep.Qcoeff$$

(where $Qstep = 12$ as before) to produce the block of coefficients shown in Table 3.13. Note that the block is significantly different from the original DCT coefficients (Table 3.10) due to

Table 3.13 Rescaled coefficients

−12	24	24	0	0	−12	0	0
12	−12	−12	0	0	12	0	0
−12	12	−12	0	0	0	0	0
−12	−12	0	0	0	0	0	0
24	−12	0	0	0	0	0	0
−12	0	0	0	0	0	0	0
0	0	0	0	0	0	0	0
0	0	0	0	0	0	0	0

Table 3.14 Decoded residual luminance samples

−3	−3	−1	1	−1	−1	−1	−3
5	3	2	0	−3	−4	−5	−6
9	6	1	−3	−5	−6	−5	−4
9	8	1	−4	−1	1	4	10
7	8	−1	−6	−1	2	5	14
2	3	−8	−15	−11	−11	−11	−2
2	5	−7	−17	−13	−16	−20	−11
12	16	3	−6	−1	−6	−11	−3

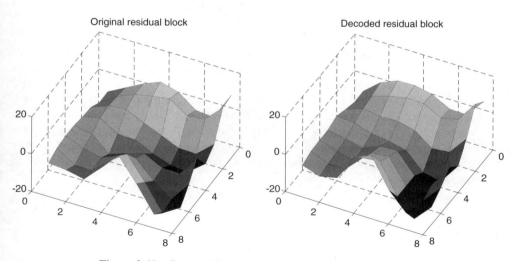

Figure 3.61 Comparison of original and decoded residual blocks

the quantisation process. An Inverse DCT is applied to create a decoded residual block (Table 3.14) which is similar but not identical to the original residual block (Table 3.9). The original and decoded residual blocks are plotted side by side in Figure 3.61 and it is clear that the decoded block has less high-frequency variation because of the loss of high-frequency DCT coefficients through quantisation.

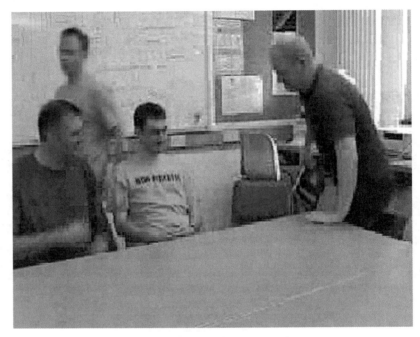

Figure 3.62   Decoded frame F'_n

The decoder forms its own predicted motion vector based on previously decoded vectors and recreates the original motion vector (0, 1). Using this vector, together with its own copy of the previously decoded frame F'_{n-1}, the decoder reconstructs the macroblock. The complete decoded frame is shown in Figure 3.62. Because of the quantisation process, some distortion has been introduced, for example around detailed areas such as the faces and the equations on the whiteboard and there are some obvious edges along 8×8 block boundaries. The complete sequence was compressed by around 300 times (i.e. the coded sequence occupies less than 1/300 the size of the uncompressed video) and so significant compression was achieved at the expense of relatively poor image quality.

3.7 CONCLUSIONS

The video coding tools described in this chapter, motion compensated prediction, transform coding, quantisation and entropy coding, form the basis of the reliable and effective coding model that has dominated the field of video compression for over 10 years. This coding model is at the heart of the two standards described in this book. The technical details of the standards are dealt with in Chapters 5 and 6 but first Chapter 4 introduces the standards themselves.

3.8 REFERENCES

1. ISO/IEC 14495-1:2000 Information technology – lossless and near-lossless compression of continuous-tone still images: Baseline, (JPEG-LS).
2. B. Horn and B. G. Schunk, Determining optical flow, *Artificial Intelligence*, **17**, 185–203, 1981.
3. K. R. Rao and P. Yip, *Discrete Cosine Transform*, Academic Press, 1990.
4. S. Mallat, *A Wavelet Tour of Signal Processing*, Academic Press, 1999.
5. N. Nasrabadi and R. King, Image coding using vector quantisation: a review, *IEEE Trans. Commun*, **36** (8), August 1988.
6. W. A. Pearlman, Trends of tree-based, set-partitioned compression techniques in still and moving image systems, *Proc. International Picture Coding Symposium*, Seoul, April 2001.
7. D. Huffman, A method for the construction of minimum redundancy codes, *Proc. of the IRE*, **40**, pp. 1098–1101, 1952.
8. I. Witten, R. Neal and J. Cleary, Arithmetic coding for data compression, *Communications of the ACM*, **30** (6), June 1987.
9. ITU-T Recommendation, Information technology – coded representation of picture and audio information – progressive bi-level image compression, T.82 (JBIG).

4

The MPEG-4 and H.264 Standards

4.1 INTRODUCTION

An understanding of the process of creating the standards can be helpful when interpreting or implementing the documents themselves. In this chapter we examine the role of the ISO MPEG and ITU VCEG groups in developing the standards. We discuss the mechanisms by which the features and parameters of the standards are chosen and the driving forces (technical and commercial) behind these mechanisms. We explain how to 'decode' the standards and extract useful information from them and give an overview of the two standards covered by this book, MPEG-4 Visual (Part 2) [1] and H.264/MPEG-4 Part 10 [2]. The technical scope and details of the standards are presented in Chapters 5 and 6 and in this chapter we concentrate on the target applications, the 'shape' of each standard and the method of specifying the standard. We briefly compare the two standards with related International Standards such as MPEG-2, H.263 and JPEG.

4.2 DEVELOPING THE STANDARDS

Creating, maintaining and updating the ISO/IEC 14496 ('MPEG-4') set of standards is the responsibility of the Moving Picture Experts Group (MPEG), a study group who develop standards for the International Standards Organisation (ISO). The emerging H.264 Recommendation (also known as MPEG-4 Part 10, 'Advanced Video Coding' and formerly known as H.26L) is a joint effort between MPEG and the Video Coding Experts Group (VCEG), a study group of the International Telecommunications Union (ITU).

MPEG developed the highly successful MPEG-1 and MPEG-2 standards for coding video and audio, now widely used for communication and storage of digital video, and is also responsible for the MPEG-7 standard and the MPEG-21 standardisation effort. VCEG was responsible for the first widely-used videotelephony standard (H.261) and its successor, H.263, and initiated the early development of the H.26L project. The two groups set-up the collaborative Joint Video Team (JVT) to finalise the H.26L proposal and convert it into an international standard (H.264/MPEG-4 Part 10) published by both ISO/IEC and ITU-T.

H.264 and MPEG-4 Video Compression: Video Coding for Next-generation Multimedia.
Iain E. G. Richardson.   © 2003 John Wiley & Sons, Ltd. ISBN: 0-470-84837-5

Table 4.1   MPEG sub-groups and responsibilities [5]

Sub-group	Responsibilities
Requirements	Identifying industry needs and requirements of new standards.
Systems	Combining audio, video and related information; carrying the combined data on delivery mechanisms.
Description	Declaring and describing digital media items.
Video	Coding of moving images.
Audio	Coding of audio.
Synthetic Natural Hybrid Coding	Coding of synthetic audio and video for integration with natural audio and video.
Integration	Conformance testing and reference software.
Test	Methods of subjective quality assessment.
Implementation	Experimental frameworks, feasibility studies, implementation guidelines.
Liaison	Relations with other relevant groups and bodies.
Joint Video Team	See below.

4.2.1 ISO MPEG

The Moving Picture Experts Group is a Working Group of the International Organisation for Standardization (ISO) and the International Electrotechnical Commission (IEC). Formally, it is Working Group 11 of Subcommittee 29 of Joint Technical Committee 1 and hence its official title is ISO/IEC JTC1/SC29/WG11.

MPEG's remit is to develop standards for compression, processing and representation of moving pictures and audio. It has been responsible for a series of important standards starting with MPEG-1 (compression of video and audio for CD playback) and following on with the very successful MPEG-2 (storage and broadcasting of 'television-quality' video and audio). MPEG-4 (coding of audio-visual objects) is the latest standard that deals specifically with audio-visual coding. Two other standardisation efforts (MPEG-7 [3] and MPEG-21 [4]) are concerned with multimedia content representation and a generic 'multimedia framework' respectively[1]. However, MPEG is arguably best known for its contribution to audio and video compression. In particular, MPEG-2 is ubiquitous at the present time in digital TV broadcasting and DVD-Video and MPEG Layer 3 audio coding ('MP3') has become a highly popular mechanism for storage and sharing of music (not without controversy).

MPEG is split into sub-groups of experts, each addressing a particular issue related to the standardisation efforts of the group (Table 4.1). The 'Experts' in MPEG are drawn from a diverse, world-wide range of companies, research organisations and institutes. Membership of MPEG and participation in its meetings is restricted to delegates of National Standards Bodies. A company or institution wishing to participate in MPEG may apply to join its National Standards Body (for example, in the UK this is done through committee IST/37 of the British Standards Institute). Joining the standards body, attending national meetings and attending the international MPEG meetings can be an expensive and time consuming exercise.

---

[1] So what happened to the missing numbers? MPEG-3 was originally intended to address coding for high-definition video but this was incorporated into the scope of MPEG-2. By this time, MPEG-4 was already under development and so '3' was missed out. The sequential numbering system was abandoned for MPEG-7 and MPEG-21.

However, the benefits include access to the private documents of MPEG (including access to the standards in draft form before their official publication, providing a potential market lead over competitors) and the opportunity to shape the development of the standards.

MPEG conducts its business at meetings that take place every 2–3 months. Since 1988 there have been 64 meetings (the first was in Ottawa, Canada and the latest in Pattaya, Thailand). A meeting typically lasts for five days, with ad hoc working groups meeting prior to the main meeting. At the meeting, proposals and developments are deliberated by each sub-group, successful contributions are approved and, if appropriate, incorporated into the standardisation process. The business of each meeting is recorded by a set of input and output documents. Input documents include reports by ad hoc working groups, proposals for additions or changes to the developing standards and statements from other bodies and output documents include meeting reports and draft versions of the standards.

4.2.2 ITU-T VCEG

The Video Coding Experts Group is a working group of the International Telecommunication Union Telecommunication Standardisation Sector (ITU-T). ITU-T develops standards (or 'Recommendations') for telecommunication and is organised in sub-groups. Sub-group 16 is responsible for multimedia services, systems and terminals, working party 3 of SG16 addresses media coding and the work item that led to the H.264 standard is described as Question 6. The official title of VCEG is therefore ITU-T SG16 Q.6. The name VCEG was adopted relatively recently and previously the working group was known by its ITU designation and/or as the Low Bitrate Coding (LBC) Experts Group.

VCEG has been responsible for a series of standards related to video communication over telecommunication networks and computer networks. The H.261 videoconferencing standard was followed by the more efficient H.263 which in turn was followed by later versions (informally known as H.263+ and H.263++) that extended the capabilities of H.263. The latest standardisation effort, previously known as 'H.26L', has led to the development and publication of Recommendation H.264.

Since 2001, this effort has been carried out cooperatively between VCEG and MPEG and the new standard, entitled 'Advanced Video Coding' (AVC), is jointly published as ITU-T H.264 and ISO/IEC MPEG-4 Part 10.

Membership of VCEG is open to any interested party (subject to approval by the chairman). VCEG meets at approximately three-monthly intervals and each meeting considers a series of objectives and proposals, some of which are incorporated into the developing draft standard. Ad hoc groups are formed in order to investigate and report back on specific problems and questions and discussion is carried out via email reflector between meetings. In contrast with MPEG, the input and output documents of VCEG (and now JVT, see below) are publicly available. Earlier VCEG documents (from 1996 to 1992) are available at [6] and documents since May 2002 are available by FTP [7].

4.2.3 JVT

The Joint Video Team consists of members of ISO/IEC JTC1/SC29/WG11 (MPEG) and ITU-T SG16 Q.6 (VCEG). JVT came about as a result of an MPEG requirement for advanced video coding tools. The core coding mechanism of MPEG-4 Visual (Part 2) is based on rather

Table 4.2   MPEG-4 and H.264 development history

1993	MPEG-4 project launched. Early results of H. 263 project produced.
1995	MPEG-4 call for proposals including efficient video coding and content-based functionalities. H.263 chosen as core video coding tool
1998	Call for proposals for H.26L.
1999	MPEG-4 Visual standard published. Initial Test Model (TM1) of H.26L defined.
2000	MPEG call for proposals for advanced video coding tools.
2001	Edition 2 of the MPEG-4 Visual standard published. H.26L adopted as basis for proposed MPEG-4 Part 10. JVT formed.
2002	Amendments 1 and 2 (Studio and Streaming Video profiles) to MPEG-4 Visual Edition 2 published. H.264 technical content frozen.
2003	H.264/MPEG-4 Part 10 ('Advanced Video Coding') published.

old technology (the original H.263 standard, published as a standard in 1995). With recent advances in processor capabilities and video coding research it was clear that there was scope for a step-change in video coding performance. After evaluating several competing technologies in June 2001, it became apparent that the H.26L test model CODEC was the best choice to meet MPEG's requirements and it was agreed that members of MPEG and VCEG would form a Joint Video Team to manage the final stages of H.26L development. Under its Terms of Reference [8], JVT meetings are co-located with meetings of VCEG and/or MPEG and the outcomes are reported back to the two parent bodies. JVT's main purpose was to see the H.264 Recommendation / MPEG-4 Part 10 standard through to publication ; now that the standard is complete, the group's focus has switched to extensions to support other colour spaces and increased sample accuracy.

4.2.4 Development History

Table 4.2 lists some of the major milestones in the development of MPEG-4 Visual and H.264. The focus of MPEG-4 was originally to provide a more flexible and efficient update to the earlier MPEG-1 and MPEG-2 standards. When it became clear in 1994 that the newly-developed H.263 standard offered the best compression technology available at the time, MPEG changed its focus and decided to embrace object-based coding and functionality as the distinctive element of the new MPEG-4 standard. Around the time at which MPEG-4 Visual was finalised (1998/99), the ITU-T study group began evaluating proposals for a new video coding initiative entitled H.26L (the L stood for 'Long Term'). The developing H.26L test model was adopted as the basis for the proposed Part 10 of MPEG-4 in 2001.

4.2.5 Deciding the Content of the Standards

Development of a standard for video coding follows a series of steps. A work plan is agreed with a set of functional and performance objectives for the new standard. In order to decide upon the basic technology of the standard, a competitive trial may be arranged in which interested parties are invited to submit their proposed solution for evaluation. In the case of video and image coding, this may involve evaluating the compression and processing performance of a number of software CODECs. In the cases of MPEG-4 Visual and H.264, the chosen core technology for video coding is based on block-based motion compensation followed by transformation and quantisation of the residual. Once the basic algorithmic core of the standard has been decided, much further work is required to develop the detail of the standard. Proposals for algorithms, methods or functionalities are made by companies or organisations with an interest in the outcome of the standardisation effort and are submitted to the standardisation group for evaluation. At each meeting of the standards group, proposal submissions are discussed and may be adopted or discarded. Through this process, the detail of the standard is gradually refined.

A software model ('reference software') is developed that implements the current agreed functionality, together with a document describing this software ('test model'). The reference software and test model are revised and updated to incorporate each agreed refinement to the standard. Eventually, the standards group decides that the test model document is at a sufficiently mature stage and it is converted into a draft standard. After further refinement and a series of ballots, the document is published as an International Standard.

The content of the final International Standard is reached through an organic development process that involves hundreds of participants (the members of the standardisation group) and many modifications and compromises. The final document is shaped by a number of driving forces. On the technical side, the contributors to the MPEG and VCEG meetings are drawn from the leading research institutes, universities and companies in the video coding field. The standards groups attempt to incorporate the best of current research and development in order to achieve the desired compression and functional performance within practical computational constraints. On the commercial side, the initial work plan of the standardisation group aims to anticipate the technology requirements of future applications. Technical proposals have to be evaluated in terms of cost-effective implementation as well as performance. Many of the companies represented at the standards group meetings have significant vested interests in the outcome of the standardisation process. In some cases, the company behind a particular proposal may own key patents that could lead to lucrative licensing revenue if the proposal is adopted as part of a worldwide standard. The JVT has asked members to identify any patented technology in proposals and is attempting to ensure that the Baseline Profile of H.264 can be implemented without royalty payments to patent holders, in order to promote widespread adoption of the standard.

On the practical side, the standards bodies work to tight timescales and it may be necessary to compromise on some aspects of the work plan in order to complete the standard on time, since it is important to ensure that the standard is technically competitive when it is released. In some ways the MPEG-4 Visual and H.264 standardisation efforts have been a race against time, since many proprietary (nonstandard) image and video coding systems and packages continue to be developed and released. Outside the constraints of the standardisation process, it is possible for a proprietary system to gain market share quickly and to make a standard effectively obsolete by the time it is published.

4.3 USING THE STANDARDS

MPEG-4 Visual (Second Edition) is a document consisting of 539 A4 pages and H.264 is over 250 A4 pages long. The standards are specified in detail in order to attempt to ensure compatibility between conforming systems and products but the level of detail in the standards documents can make it difficult to extract information from them. In this section we give an overview of the content of the standards and suggest how to interpret and apply them.

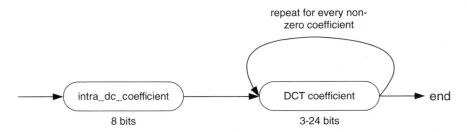

Figure 4.1 MPEG-4 Visual block syntax

4.3.1 What the Standards Cover

In common with earlier video coding standards, the MPEG-4 Visual and H.264 standards do not specify a video encoder. Instead, the standards specify the *syntax* of a coded bitstream (the binary codes and values that make up a conforming bitstream), the *semantics* of these syntax elements (what they mean) and the *process* by which the syntax elements may be decoded to produce visual information.

> ### *Example:* **MPEG-4 Visual, coefficient block syntax**
>
> A block of quantised and run-level coded DCT coefficients is represented in an MPEG-4 Visual bitstream using the syntax shown in Figure 4.1 (Simple Profile / Short Header mode). The meaning (semantics) of the syntax elements is as follows.
>
> intra_dc_coefficient: A fixed-length code defining the value of the DC (0,0) coefficient.
> DCT coefficient: A variable-length code defining the values of RUN, LEVEL and LAST. RUN is the number of zeros preceding each nonzero quantised coefficient; LEVEL is the value of the nonzero coefficient; and LAST is a flag indicating whether this is the final nonzero coefficient in the block.
>
> The standard describes how the actual values of these data elements (coefficients, zero runs and LAST) can be decoded from the fixed-length or variable-length codes.

By specifying syntax and semantics of the coded bitstream, the standard defines what is acceptable as a 'compliant bitstream'. A compliant encoder has to produce a bitstream that is correctly decodeable using the decoding process defined in the standard. As well as specifying syntax, semantics and decoding process, both of the standards define a *hypothetical reference decoder* that, together with certain defined *levels* of operation, places practical limits on the encoder. These limits include, for example, the maximum coded bitrate and maximum picture size that an encoder may produce and are important because they constrain the memory and processing requirements of a decoder.

4.3.2 Decoding the Standards

The MPEG-4 Visual and H.264 standards are not written as development guides but rather with the goal of ensuring compatibility between CODECs developed by different manufacturers. For example, an H.264-compliant bitstream generated by any encoder should be decodeable by any H.264-compliant decoder. Ensuring that this takes place is a difficult task and a potential drawback of this overriding goal is that the standards documents are long and complex and can be difficult to follow.

It is possible to work out how to design a decoder by reading the standard, since the decoding process is specified in some detail. The encoding process is not specified at all, in order to give designers the flexibility to choose their own method of encoding. However, the syntax and decoding process places certain constraints on the encoder design. For example, in order to produce a bitstream that is compatible with MPEG-4 Simple Profile, it is necessary for an encoder to carry out the functions described in Chapter 3, Section 3.6 or functions that achieve an equivalent result. It is therefore useful to have a working knowledge of at least one design of CODEC that can meet the requirements of the standard.

Video coding is an important component of the multimedia entertainment industry, however, the video coding standards can be difficult to interpret by nonexperts. This has led to a plethora of tutorials, training courses, on-line resources and books (including this one) that attempt to help explain and de-mystify the standards. Many of these resources are excellent but they should be used with caution by developers of standards-compliant video coding applications, since the 'correct' interpretation is always the standard itself.

A useful approach to understanding and 'decoding' the standards is to become familiar with a reference implementation that complies with the standard. Reference software is available for both MPEG-4 Visual and H.264 and contains source code for an encoder that produces a compliant bitstream and a decoder that conforms to the decoding process defined in the standard.

4.3.3 Conforming to the Standards

For an *encoder*, conforming to a video coding standard means that the encoder produces a bitstream that meets the requirements of the specified syntax and is capable of being decoded using the decoding process described in the standard, whilst meeting constraints on parameters such as coded bitrate. For a *decoder*, conforming to the standard means that the decoder can decode a conforming bitstream whilst meeting any specified constraints and accuracy parameters. Conformance is defined further by *profiles* (agreed subsets of the syntax and tools available within the standard) and *levels* (agreed performance limits).

A good guide to conformance is interoperability with reference software (software implementations of MPEG-4 Visual and H.264 that are publicly available for developers to download). If a third-party encoder or decoder can successfully interwork with the reference software decoder or encoder, it indicates that the CODEC *may* conform to the standard. However, this does not necessarily prove conformance. The reference software models encode and decode video 'offline' and the input and output is file-based, making it difficult to test directly the operation of a real-time application which may, for example, require 'live' video data to be encoded and transmitted across a network. Furthermore, there have been known mismatches between the reference software models and the specifications within the standards documents[2] (though the Experts Groups try to minimise this type of problem by correcting errors and mismatches). Finally, some compatibility problems or errors may only become apparent under certain operating conditions or with certain types of video material.

[2] At the time of writing, the encoder part of the H.264 reference software does not comply with the final version of the standard.

Interoperability testing (exchanging coded bitstreams with developers of other encoders and decoders) is considered to be a valuable way of testing and identifying a wide range of potential compatibility problems. Industry bodies such as the Internet Streaming Media Alliance (ISMA) [10] and the MPEG-4 Industry Forum (M4IF) [11] actively promote and organise interoperability tests between manufacturers.

In MPEG-4, testing conformance of bitstreams and decoders is dealt with in Part 4 (Conformance Testing). A bitstream is deemed to conform to the standard if it includes the correct syntax elements for a particular Profile and does not exceed (or cause a decoder to exceed) the limits imposed by the operating Level. A decoder conforms to the standard if it can decode any bitstream that conforms to the Profile/Level combination. Some conformance test bitstreams are included as an Annex to MPEG-4 Part 4.

4.4 OVERVIEW OF MPEG-4 VISUAL/PART 2

MPEG-4 Visual (Part 2 of ISO/IEC 14496, 'Coding of Audio-Visual Objects') is a large document that covers a very wide range of functionalities, all related to the coding and representation of visual information. The standard deals with the following types of data, among others:

- moving video (rectangular frames);
- video objects (arbitrary-shaped regions of moving video);
- 2D and 3D mesh objects (representing deformable objects);
- animated human faces and bodies;
- static texture (still images).

The standard specifies a set of coding 'tools' that are designed to represent these data types in compressed (coded) form. With a diverse set of tools and supported data types, the MPEG-4 Visual standard can support many different applications, including (but by no means limited to) the following:

- 'legacy' video applications such as digital TV broadcasting, videoconferencing and video storage;
- 'object-based' video applications in which a video scene may be composed of a number of distinct video objects, each independently coded;
- rendered computer graphics using 2D and 3D deformable mesh geometry and/or animated human faces and bodies;
- 'hybrid' video applications combining real-world (natural) video, still images and computer-generated graphics;
- streaming video over the Internet and mobile channels;
- high-quality video editing and distribution for the studio production environment.

Despite the bewildering range of coding tools specified by the standard, at the heart of MPEG-4 Visual is a rather simple video coding mechanism, a block-based video CODEC that uses motion compensation followed by DCT, quantisation and entropy coding (essentially the DPCM/DCT model described in Chapter 3). The syntax of this core CODEC is (under certain constraints) identical to that of baseline H.263. Most of the other functions and tools supported by the standard are obtained by adding features to this core, whilst the tools for coding meshes, still images and face/body animation parameters are developed separately. The technical detail contained within the standard is covered in Chapter 5. As a guide to approaching the standard, it is useful to examine the main sections of the document itself and what they cover.

The *Introduction* gives an overview of some of the target applications, approaches and data types, with a particular emphasis on 2D and 3D mesh objects (the reason for this emphasis is not clear).

Sections 1 to 5 are a preamble to the technical detail of the standard. Section 1 describes the scope of the standard, Section 2 references other standards documents, Section 3 contains a useful (but not complete) list of terminology and definitions, Section 4 lists standard symbols and abbreviations and Section 5 explains the conventions for describing the syntax of the standard.

Section 6 describes the syntax and semantics of MPEG-4 Visual. The various structure elements in an MPEG-4 Visual bitstream are described, together with the assumed format for uncompressed video. The complete syntax of a compliant bitstream is defined and this section specifies which syntax elements are part of the standard and the sequence(s) in which they are allowed to occur. Where necessary, the semantics (meaning and allowed values) of each syntax element are described. Section 6 defines the acceptable parameters of an MPEG-4 Visual bitstream and a compliant encoder has to produce a bitstream that follows this syntax.

Section 7 describes a set of processes for decoding an MPEG-4 Visual bitstream. This section defines the series of steps required to decode a compliant bitstream and convert it to a visual scene or visual object. First, the basic steps of decoding residual data ('texture') and shape parameters and motion compensation are described, followed by the special cases required for the other coding tools (interlaced video decoding, still image and 'sprite' decoding, scalable decoding, etc.). Hence Section 7 defines how an MPEG-4 Visual bitstream should be decoded. A compliant decoder has to follow these *or equivalent* processes.

Section 8 discusses how separately-decoded visual 'objects' should be composed to make a visual scene and *Section 9* defines a set of conformance points known as 'profiles' and 'levels'. These specify subsets of the coding tools (profiles) and performance limits (levels) that CODECs may choose to adopt. These conformance points are particularly important for developers and manufacturers since it is unlikely that any one application will require all of the various tools supported by the standard.

The remainder of the document (another 200 pages) consists of 15 *Annexes* (Annex A to Annex O). Some of these are *normative*, i.e. they are considered to be an essential part of the text. Examples include definitions of the Discrete Cosine and Discrete Wavelet Transforms used in the standard, tables of variable-length codes used in the MPEG-4 Visual bitstream and the definition of a Video Buffering Verifier that places limits on CODEC performance. The remainder are *informative* Annexes, i.e. they may help clarify the standard but are not intended to be essential definitions. Examples include a description of error-resilience features, suggested mechanisms for pre- and post-processing video, suggested bitrate control algorithms and a list of companies who own patents that may cover aspects of the standard.

4.5 OVERVIEW OF H.264 / MPEG-4 PART 10

H.264 has a narrower scope than MPEG-4 Visual and is designed primarily to support efficient and robust coding and transport of rectangular video frames. Its original aim was to provide similar functionality to earlier standards such as H.263+ and MPEG-4 Visual (Simple Profile) but with significantly better compression performance and improved support for reliable transmission. Target applications include two-way video communication (videoconferencing or videotelephony), coding for broadcast and high quality video and video streaming over packet networks. Support for robust transmission over networks is built in and the standard is designed to facilitate implementation on as wide a range of processor platforms as possible.

The standard is specified in more detail than many earlier standards (including MPEG-4 Visual) in an attempt to minimise the possibility for misinterpretation by developers. The detailed level of specification means the standard document is relatively long and makes some sections of the standard difficult to follow.

The *Introduction* lists some target applications, explains the concept of Profiles and Levels and gives a brief overview of the coded representation.

Sections 1 to 5 are a preamble to the detail and include references to other standards documents, a list of terminology and definitions, standard abbreviations, arithmetic operators and mathematical functions.

Section 6 describes the input and output data formats. The standard supports coding of 4:2:0 progressive and interlaced video and the sampling formats assumed by the standard are described here, together with the format of the coded bitstream and the order of processing of video frames. This section also defines methods for finding 'neighbours' of a coded element, necessary for the decoding process.

Section 7 describes the syntax and semantics. The structure and sequence of each syntax element are described in a similar way to MPEG-4 Visual. The semantics (meaning and allowed values) of each syntax element are defined.

Section 8 describes the processes involved in decoding slices (the basic coded unit of H.264). The first subsections deal with higher-level issues such as picture boundary detection and reference picture management. This is followed by intra- and inter- prediction, transform coefficient decoding, reconstruction and the built-in deblocking filter.

Section 9 describes how a coded bitstream should be 'parsed' (i.e. how syntax elements should be extracted from the bitstream). There are two main methods, one based on variable-length codes and one based on context-adaptive binary arithmetic codes.

Annex A defines the three profiles supported by H.264 (Baseline, Main and Extended) and specified performance points ('levels').

Annex B defines the format of a byte stream (a series of coded data units suitable for transmission over a circuit-switched network connection).

Annex C defines a Hypothetical Reference Decoder. In a similar way to the MPEG-4 Visual 'Video Buffering Verifier', this places limits on coded bitrate and decoded picture rate.

Annex D and *Annex E* describe Supplemental Enhancement Information (SEI), and Video Usability Information (VUI), 'side information' that may be used by a decoder to enhance or improve the decoded display but which is not essential to correct decoding of a sequence.

4.6 COMPARISON OF MPEG-4 VISUAL AND H.264

Table 4.3 summarises some of the main differences between the two standards. This table is by no means a complete comparison but it highlights some of the important differences in approach between MPEG-4 Visual and H.264.

Table 4.3 Summary of differences between MPEG-4 Visual and H.264

Comparison	MPEG-4 Visual	H.264
Supported data types	Rectangular video frames and fields, arbitrary-shaped video objects, still texture and sprites, synthetic or synthetic–natural hybrid video objects, 2D and 3D mesh objects	Rectangular video frames and fields
Number of profiles	19	3
Compression efficiency	Medium	High
Support for video streaming	Scalable coding	Switching slices
Motion compensation minimum block size	8×8	4×4
Motion vector accuracy	half or quarter-pixel	quarter-pixel
Transform	8×8 DCT	4×4 DCT approximation
Built-in deblocking filter	No	Yes
Licence payments required for commercial implementation	Yes	Probably not (baseline profile); probably (main and extended profiles)

4.7 RELATED STANDARDS

4.7.1 JPEG and JPEG2000

JPEG (the Joint Photographic Experts Group, an ISO working group similar to MPEG) has been responsible for a number of standards for coding of still images. The most relevant of these for our topic are the JPEG standard [12] and the JPEG2000 standard [13]. Each share features with MPEG-4 Visual and/or H.264 and whilst they are intended for compression of still images, the JPEG standards have made some impact on the coding of moving images. The 'original' JPEG standard supports compression of still photographic images using the 8×8 DCT followed by quantisation, reordering, run-level coding and variable-length entropy coding and so has similarities to MPEG-4 Visual (when used in Intra-coding mode). JPEG2000 was developed to provide a more efficient successor to the original JPEG. It uses the Discrete Wavelet Transform as its basic coding method and hence has similarities to the Still Texture coding tools of MPEG-4 Visual. JPEG2000 provides superior compression performance to JPEG and does not exhibit the characteristic blocking artefacts of a DCT-based compression method. Despite the fact that JPEG is 'old' technology and is outperformed by JPEG2000 and many proprietary compression formats, it is still very widely used for storage of images in digital cameras, PCs and on web pages. Motion JPEG (a nonstandard method of compressing a sequence of video frames using JPEG) is popular for applications such as video capture, PC-based video editing and security surveillance.

4.7.2 MPEG-1 and MPEG-2

The first MPEG standard, MPEG-1 [14], was developed for the specific application of video storage and playback on Compact Disks. A CD plays for around 70 minutes (at 'normal' speed) with a transfer rate of 1.4 Mbit/s. MPEG-1 was conceived to support the video CD, a format for consumer video storage and playback that was intended to compete with VHS videocassettes. MPEG-1 Video uses block-based motion compensation, DCT and quantisation (the hybrid DPCM/DCT CODEC model described in Chapter 3) and is optimised for a compressed video bitrate of around 1.2 Mbit/s. Alongside MPEG-1 Video, the Audio and Systems parts of the standard were developed to support audio compression and creation of a multiplexed bitstream respectively. The video CD format did not take off commercially, perhaps because the video quality was not sufficiently better than VHS tape to encourage consumers to switch to the new technology. MPEG-1 video is still widely used for PC- and web-based storage of compressed video files.

Following on from MPEG-1, the MPEG-2 standard [15] aimed to support a large potential market, digital broadcasting of compressed television. It is based on MPEG-1 but with several significant changes to support the target application including support for efficient coding of interlaced (as well as progressive) video, a more flexible syntax, some improvements to coding efficiency and a significantly more flexible and powerful 'systems' part of the standard. MPEG-2 was the first standard to introduce the concepts of Profiles and Levels (defined conformance points and performance limits) as a way of encouraging interoperability without restricting the flexibility of the standard.

MPEG-2 was (and still is) a great success, with worldwide adoption for digital TV broadcasting via cable, satellite and terrestrial channels. It provides the video coding element of DVD-Video, which is succeeding in finally replacing VHS videotape (where MPEG-1 and the video CD failed). At the time of writing, many developers of legacy MPEG-1 and MPEG-2 applications are looking for the coding standard for the next generation of products, ideally providing better compression performance and more robust adaptation to networked transmission than the earlier standards. Both MPEG-4 Part 2 (now relatively mature and with all the flexibility a developer could ever want) and Part 10/H.264 (only just finalised but offering impressive performance) are contenders for a number of key current- and next-generation applications.

4.7.3 H.261 and H.263

H.261 [16] was the first widely-used standard for videoconferencing, developed by the ITU-T to support videotelephony and videoconferencing over ISDN circuit-switched networks. These networks operate at multiples of 64 kbit/s and the standard was designed to offer computationally-simple video coding for these bitrates. The standard uses the familiar hybrid DPCM/DCT model with integer-accuracy motion compensation.

In an attempt to improve on the compression performance of H.261, the ITU-T working group developed H.263 [17]. This provides better compression than H.261, supporting basic video quality at bitrates of below 30 kbit/s, and is part of a suite of standards designed to operate over a wide range of circuit- and packet-switched networks. The 'baseline' H.263 coding model (hybrid DPCM/DCT with half-pixel motion compensation) was adopted as the core of MPEG-4 Visual's Simple Profile (see Chapter 5). The original version of H.263 includes four optional coding modes (each described in an Annex to the standard) and Issue 2 added a series of further optional modes to support features such as improved compression efficiency and robust transmission over lossy networks. The terms 'H.263+' and 'H.263++' are used to describe CODECs that support some or all of the optional coding modes. The choice of supported modes is up to the CODEC manufacturer and the wide range of possible mode combinations (and the corresponding difficulty of ensuring compatibility between CODECs) was one motivation for the relatively simple structure of H.264's three profiles.

4.7.4 Other Parts of MPEG-4

There are currently 16 Parts of MPEG-4 either published or under development. Parts 1–8 and 10 are published International Standards or Technical Reports whereas the remaining Parts are not yet published in final form.

Part 1, Systems: Scene description, multiplexing of audio, video and related information, synchronisation, buffer management, intellectual property management.

Part 2, Visual: Coding of 'natural' and 'synthetic' visual objects (see Chapter 5).

Part 3, Audio: Coding of natural and synthetic audio objects.

Part 4, Conformance Testing: Conformance conditions, test procedures, test bitstreams.

Part 5, Reference Software: Publicly-available software that implements most tools in the Standard.

Part 6, Delivery Multimedia Integration Framework: A session protocol for multimedia streaming.

Part 7, Optimised Visual Reference Software: Optimised software implementation of selected Visual coding tools. This Part is a Technical Report (and not an International Standard).

Part 8, Carriage of MPEG-4 over IP: Specifies the mechanism for carrying MPEG-4 coded data over Internet Protocol (IP) networks.

Part 9, Reference Hardware Description: VHDL descriptions of MPEG-4 coding tools (suitable for implementation in ICs). This Part is a Technical Report and is still under development.

Part 10, Advanced Video Coding: Efficient coding of natural video (see Chapter 6). This Part is under development and due to become an International Standard in 2003.

Part 11, Scene Description and Application Engine.

Part 12, ISO Base Media File Format.

Part 13, Intellectual Property Management and Protection Extensions.

Part 14, MPEG-4 File Format (see Chapter 7).

Part 15, AVC File Format (see Chapter 7).

Part 16, Animation Framework Extension.

4.8 CONCLUSIONS

In this chapter we have set the scene for the two standards, H.264 and MPEG-4 Visual, charting their development history and summarising their main features and contents. For most users of the standards, the key question is 'how do I encode and decode video ?' and this question will be addressed in Chapter 5 (covering the detail of MPEG-4 Visual) and Chapter 6 (concentrating on H.264).

4.9 REFERENCES

1. ISO/IEC 14496-2, Coding of audio-visual objects – Part 2: Visual, 2001.
2. ISO/IEC 14496-10 and ITU-T Rec. H.264, Advanced Video Coding, 2003.
3. ISO/IEC 15938, Information technology – multimedia content description interface (MPEG-7), 2002.
4. ISO/IEC 21000, Information technology – multimedia framework (MPEG-21), 2003.
5. Terms of Reference, MPEG home page, http://mpeg.telecomitalialab.com/.
6. VCEG document site, http://standards.pictel.com/ftp/video-site/.
7. JVT experts FTP site, ftp://ftp.imtc-files.org/jvt-experts/.
8. Terms of Reference for Joint Video Team Activities, http://www.itu.int/ITU-T/studygroups/com16/jvt/.
9. F. Pereira and T. Ebrahimi (eds), "The MPEG-4 Book", Prentice Hall 2002 (section 1.1)
10. Internet Streaming Media Alliance, http://www.isma.org.
11. MPEG-4 Industry Forum, http://www.m4if.org.
12. ISO/IEC 10918-1 / ITU-T Recommendation T.81, Digital compression and coding of continuous-tone still images, 1992 (JPEG).
13. ISO/IEC 15444, Information technology – JPEG 2000 image coding system, 2000.
14. ISO/IEC 11172, Information technology – coding of moving pictures and associated audio for digital storage media at up to about 1.5 Mbit/s, 1993 (MPEG-1).
15. ISO/IEC 13818, Information technology: generic coding of moving pictures and associated audio information, 1995 (MPEG-2).
16. ITU-T Recommendation H.261, Video CODEC for audiovisual services at px64 kbit/s, 1993.
17. ITU-T Recommendation H.263, Video coding for low bit rate communication, Version 2, 1998.

5

MPEG-4 Visual

5.1 INTRODUCTION

ISO/IEC Standard 14496 Part 2 [1] (MPEG-4 Visual) improves on the popular MPEG-2 standard both in terms of compression efficiency (better compression for the same visual quality) and flexibility (enabling a much wider range of applications). It achieves this in two main ways, by making use of more advanced compression algorithms and by providing an extensive set of 'tools' for coding and manipulating digital media. MPEG-4 Visual consists of a 'core' video encoder/decoder model together with a number of additional coding tools. The core model is based on the well-known hybrid DPCM/DCT coding model (see Chapter 3) and the basic function of the core is extended by tools supporting (among other things) enhanced compression efficiency, reliable transmission, coding of separate shapes or 'objects' in a visual scene, mesh-based compression and animation of face or body models.

It is unlikely that any single application would require all of the tools available in the MPEG-4 Visual framework and so the standard describes a series of *profiles*, recommended sets or groupings of tools for particular types of application. Examples of profiles include Simple (a minimal set of tools for low-complexity applications), Core and Main (with tools for coding multiple arbitrarily-shaped video objects), Advanced Real Time Simple (with tools for error-resilient transmission with low delay) and Advanced Simple (providing improved compression at the expense of increased complexity).

MPEG-4 Visual is embodied in ISO/IEC 14496-2, a highly detailed document running to over 500 pages. Version 1 was released in 1998 and further tools and profiles were added in two Amendments to the standard culminating in Version 2 in late 2001. More tools and profiles are planned for future Amendments or Versions but the 'toolkit' structure of MPEG-4 means that any later versions of 14496-2 should remain backwards compatible with Version 1.

This chapter is a guide to the tools and features of MPEG-4 Visual. Practical implementations of MPEG-4 Visual are based on one or more of the profiles defined in the standard and so this chapter is organised according to profiles. After an overview of the standard and its approach and features, the profiles for coding rectangular video frames are discussed (Simple,

H.264 and MPEG-4 Video Compression: Video Coding for Next-generation Multimedia.
Iain E. G. Richardson.   © 2003 John Wiley & Sons, Ltd. ISBN: 0-470-84837-5

Advanced Simple and Advanced Real-Time Simple profiles). These are by far the most popular profiles in use at the present time and so they are covered in some detail. Tools and profiles for coding of arbitrary-shaped objects are discussed next (the Core, Main and related profiles), followed by profiles for scalable coding, still texture coding and high-quality ('studio') coding of video.

In addition to tools for coding of 'natural' (real-world) video material, MPEG-4 Visual defines a set of profiles for coding of 'synthetic' (computer-generated) visual objects such as 2D and 3D meshes and animated face and body models. The focus of this book is very much on coding of natural video and so these profiles are introduced only briefly. Coding tools in the MPEG-4 Visual standard that are not included in any Profile (such as Overlapped Block Motion Compensation, OBMC) are (perhaps contentiously!) not covered in this chapter.

5.2 OVERVIEW OF MPEG-4 VISUAL (NATURAL VIDEO CODING)

5.2.1 Features

MPEG-4 Visual attempts to satisfy the requirements of a wide range of visual communication applications through a toolkit-based approach to coding of visual information. Some of the key features that distinguish MPEG-4 Visual from previous visual coding standards include:

- Efficient compression of progressive and interlaced 'natural' video sequences (compression of sequences of rectangular video frames). The core compression tools are based on the ITU-T H.263 standard and can out-perform MPEG-1 and MPEG-2 video compression. Optional additional tools further improve compression efficiency.
- Coding of video objects (irregular-shaped regions of a video scene). This is a new concept for standard-based video coding and enables (for example) independent coding of foreground and background objects in a video scene.
- Support for effective transmission over practical networks. Error resilience tools help a decoder to recover from transmission errors and maintain a successful video connection in an error-prone network environment and scalable coding tools can help to support flexible transmission at a range of coded bitrates.
- Coding of still 'texture' (image data). This means, for example, that still images can be coded and transmitted within the same framework as moving video sequences. Texture coding tools may also be useful in conjunction with animation-based rendering.
- Coding of animated visual objects such as 2D and 3D polygonal meshes, animated faces and animated human bodies.
- Coding for specialist applications such as 'studio' quality video. In this type of application, visual quality is perhaps more important than high compression.

5.2.2 Tools, Objects, Profiles and Levels

MPEG-4 Visual provides its coding functions through a combination of *tools, objects* and *profiles*. A *tool* is a subset of coding functions to support a specific feature (for example, basic

Table 5.1 MPEG-4 Visual profiles for coding natural video

MPEG-4 Visual profile	Main features
Simple	Low-complexity coding of rectangular video frames
Advanced Simple	Coding rectangular frames with improved efficiency and support for interlaced video
Advanced Real-Time Simple	Coding rectangular frames for real-time streaming
Core	Basic coding of arbitrary-shaped video objects
Main	Feature-rich coding of video objects
Advanced Coding Efficiency	Highly efficient coding of video objects
N-Bit	Coding of video objects with sample resolutions other than 8 bits
Simple Scalable	Scalable coding of rectangular video frames
Fine Granular Scalability	Advanced scalable coding of rectangular frames
Core Scalable	Scalable coding of video objects
Scalable Texture	Scalable coding of still texture
Advanced Scalable Texture	Scalable still texture with improved efficiency and object-based features
Advanced Core	Combines features of Simple, Core and Advanced Scalable Texture Profiles
Simple Studio	Object-based coding of high quality video sequences
Core Studio	Object-based coding of high quality video with improved compression efficiency.

Table 5.2 MPEG-4 Visual profiles for coding synthetic or hybrid video

MPEG-4 Visual profile	Main features
Basic Animated Texture	2D mesh coding with still texture
Simple Face Animation	Animated human face models
Simple Face and Body Animation	Animated face and body models
Hybrid	Combines features of Simple, Core, Basic Animated Texture and Simple Face Animation profiles

video coding, interlaced video, coding object shapes, etc.). An *object* is a video element (e.g. a sequence of rectangular frames, a sequence of arbitrary-shaped regions, a still image) that is coded using one or more tools. For example, a simple video object is coded using a limited subset of tools for rectangular video frame sequences, a core video object is coded using tools for arbitrarily-shaped objects and so on. A *profile* is a set of object types that a CODEC is expected to be capable of handling.

The MPEG-4 Visual profiles for coding 'natural' video scenes are listed in Table 5.1 and these range from Simple Profile (coding of rectangular video frames) through profiles for arbitrary-shaped and scalable object coding to profiles for coding of studio-quality video. Table 5.2 lists the profiles for coding 'synthetic' video (animated meshes or face/body models) and the hybrid profile (incorporates features from synthetic and natural video coding). These profiles are not (at present) used for natural video compression and so are not covered in detail in this book.

Profile / Object types	Simple	Advanced Simple	Advanced Real-Time Simple	Core	Main	Advanced Coding Efficiency	N-bit	Simple Scalable	Fine Granular Scalability	Core Scalable	Scalable Texture	Advanced Scalable Texture	Simple Studio	Core Studio	Simple Face Animation	Simple Face and Body Animation	Basic Animated Texture	Animated 2D Mesh
Simple	✓																	
Advanced Simple	✓	✓																
Advanced Real-Time Simple	✓		✓															
Core	✓			✓														
Advanced Core	✓			✓								✓						
Main	✓			✓	✓						✓							
Advanced Coding Efficiency	✓			✓		✓												
N-bit	✓			✓			✓											
Simple Scalable	✓							✓										
Fine Granular Scalability	✓	✓							✓									
Core Scalable	✓			✓				✓		✓								
Scalable Texture											✓							
Advanced Scalable Texture												✓						
Simple Studio													✓					
Core Studio													✓	✓				
Basic Animated Texture												✓					✓	
Simple Face Animation															✓			
Simple FBA																✓		
Hybrid	✓			✓								✓			✓		✓	✓

Figure 5.1   MPEG-4 Visual profiles and objects

Figure 5.1 lists each of the MPEG-4 Visual profiles (left-hand column) and visual object types (top row). The table entries indicate which object types are contained within each profile. For example, a CODEC compatible with Simple Profile must be capable of coding and decoding Simple objects and a Core Profile CODEC must be capable of coding and decoding Simple *and* Core objects.

Profiles are an important mechanism for encouraging interoperability between CODECs from different manufacturers. The MPEG-4 Visual standard describes a diverse range of coding tools and it is unlikely that any commercial CODEC would require the implementation of all the tools. Instead, a CODEC designer chooses a profile that contains adequate tools for the target application. For example, a basic CODEC implemented on a low-power processor may use Simple profile, a CODEC for streaming video applications may choose Advanced Real Time Simple and so on. To date, some profiles have had more of an impact on the marketplace than others. The Simple and Advanced Simple profiles are particularly popular with manufacturers and users whereas the profiles for the coding of arbitrary-shaped objects have had very limited commercial impact (see Chapter 8 for further discussion of the commercial impact of MPEG-4 Profiles).

Profiles define a subset of coding tools and *Levels* define constraints on the parameters of the bitstream. Table 5.3 lists the Levels for the popular Simple-based profiles (Simple,

Table 5.3   Levels for Simple-based profiles

Profile	Level	Typical resolution	Max. bitrate	Max. objects
Simple	L0	176 × 144	64 kbps	1 simple
	L1	176 × 144	64 kbps	4 simple
	L2	352 × 288	128 kbps	4 simple
	L3	352 × 288	384 kbps	4 simple
Advanced Simple (AS)	L0	176 × 144	128 kbps	1 AS or simple
	L1	176 × 144	128 kbps	4 AS or simple
	L2	352 × 288	384 kbps	4 AS or simple
	L3	352 × 288	768 kbps	4 AS or simple
	L4	352 × 576	3 Mbps	4 AS or simple
	L5	720 × 576	8 Mbps	4 AS or simple
Advanced Real-Time Simple (ARTS)	L1	176 × 144	64 kbps	4 ARTS or simple
	L2	352 × 288	128 kbps	4 ARTS or simple
	L3	352 × 288	384 kbps	4 ARTS or simple
	L4	352 × 288	2 Mbps	16 ARTS or simple

Advanced Simple and Advanced Real Time Simple). Each Level places constraints on the maximum performance required to decode an MPEG-4 coded sequence. For example, a multimedia terminal with limited processing capabilities and a small amount of memory may only support Simple Profile @ Level 0 bitstream decoding. The Level definitions place restrictions on the amount of buffer memory, the decoded frame size and processing rate (in macroblocks per second) and the number of video objects (one in this case, a single rectangular frame). A terminal that can cope with these parameters is guaranteed to be capable of successfully decoding any conforming Simple Profile @ Level 0 bitstream. Higher Levels of Simple Profile require a decoder to handle up to four Simple Profile video objects (for example, up to four rectangular objects covering the QCIF or CIF display resolution).

5.2.3 Video Objects

One of the key contributions of MPEG-4 Visual is a move away from the 'traditional' view of a video sequence as being merely a collection of rectangular frames of video. Instead, MPEG-4 Visual treats a video sequence as a collection of one or more *video objects*. MPEG-4 Visual defines a video object as a flexible 'entity that a user is allowed to access (seek, browse) and manipulate (cut and paste)' [1]. A video object (VO) is an area of the video scene that may occupy an arbitrarily-shaped region and may exist for an arbitrary length of time. An instance of a VO at a particular point in time is a *video object plane* (VOP).

This definition encompasses the traditional approach of coding complete frames, in which each VOP is a single frame of video and a sequence of frames forms a VO (for example, Figure 5.2 shows a VO consisting of three rectangular VOPs). However, the introduction of the VO concept allows more flexible options for coding video. Figure 5.3 shows a VO that consists of three irregular-shaped VOPs, each one existing within a frame and each coded separately (*object-based coding*).

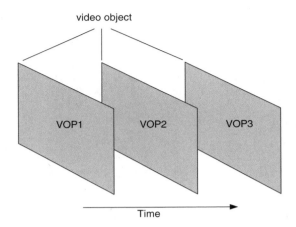

Figure 5.2 VOPs and VO (rectangular)

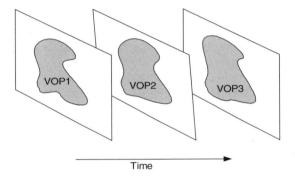

Figure 5.3 VOPs and VO (arbitrary shape)

A video scene (e.g. Figure 5.4) may be made up of a background object (VO3 in this example) and a number of separate foreground objects (VO1, VO2). This approach is potentially much more flexible than the fixed, rectangular frame structure of earlier standards. The separate objects may be coded with different visual qualities and temporal resolutions to reflect their 'importance' to the final scene, objects from multiple sources (including synthetic and 'natural' objects) may be combined in a single scene and the composition and behaviour of the scene may be manipulated by an end-user in highly interactive applications. Figure 5.5 shows a new video scene formed by adding VO1 from Figure 5.4, a new VO2 and a new background VO. Each object is coded separately using MPEG-4 Visual (the compositing of visual and audio objects is assumed to be handled separately, for example by MPEG-4 Systems [2]).

5.3 CODING RECTANGULAR FRAMES

Notwithstanding the potential flexibility offered by object-based coding, the most popular application of MPEG-4 Visual is to encode complete frames of video. The tools required

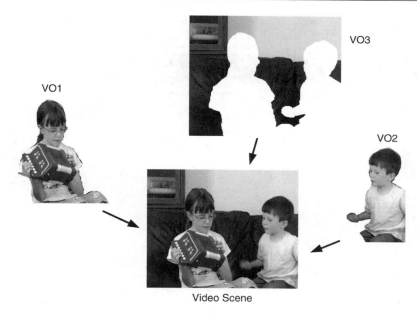

Figure 5.4 Video scene consisting of three VOs

Figure 5.5 Video scene composed of VOs from separate sources

to handle rectangular VOPs (typically complete video frames) are grouped together in the so-called *simple* profiles. The tools and objects for coding rectangular frames are shown in Figure 5.6. The basic tools are similar to those adopted by previous video coding standards, DCT-based coding of macroblocks with motion compensated prediction. The Simple profile is based around the well-known hybrid DPCM/DCT model (see Chapter 3, Section 3.6) with

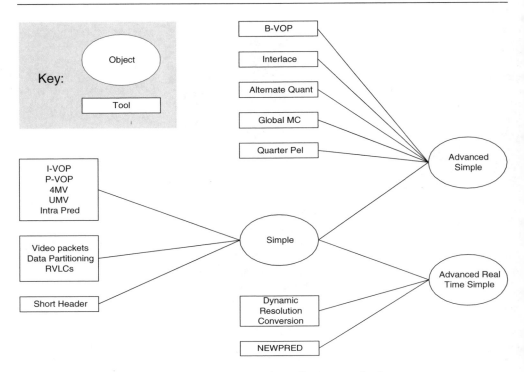

Figure 5.6   Tools and objects for coding rectangular frames

additional tools to improve coding efficiency and transmission efficiency. Because of the widespread popularity of Simple profile, enhanced profiles for rectangular VOPs have been developed. The Advanced Simple profile improves further coding efficiency and adds support for interlaced video and the Advanced Real-Time Simple profile adds tools that are useful for real-time video streaming applications.

5.3.1   Input and output video format

The input to an MPEG-4 Visual encoder and the output of a decoder is a video sequence in 4:2:0, 4:2:2 or 4:4:4 progressive or interlaced format (see Chapter 2). MPEG-4 Visual uses the sampling arrangement shown in Figure 2.11 for progressive sampled frames and the method shown in Figure 2.12 for allocating luma and chroma samples to each pair of fields in an interlaced sequence.

5.3.2   The Simple Profile

A CODEC that is compatible with Simple Profile should be capable of encoding and decoding Simple Video Objects using the following tools:

- I-VOP (Intra-coded rectangular VOP, progressive video format);
- P-VOP (Inter-coded rectangular VOP, progressive video format);

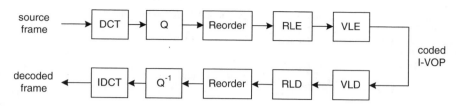

Figure 5.7 I-VOP encoding and decoding stages

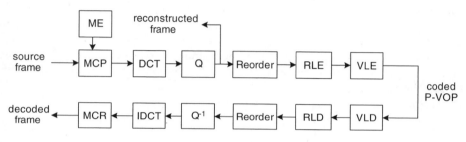

Figure 5.8 P-VOP encoding and decoding stages

- short header (mode for compatibility with H.263 CODECs);
- compression efficiency tools (four motion vectors per macroblock, unrestricted motion vectors, Intra prediction);
- transmission efficiency tools (video packets, Data Partitioning, Reversible Variable Length Codes).

5.3.2.1 The Very Low Bit Rate Video Core

The Simple Profile of MPEG-4 Visual uses a CODEC model known as the Very Low Bit Rate Video (VLBV) Core (the hybrid DPCM/DCT model described in Chapter 3). In common with other standards, the architecture of the encoder and decoder are not specified in MPEG-4 Visual but a practical implementation will require to carry out the functions shown in Figure 5.7 (coding of Intra VOPs) and Figure 5.8 (coding of Inter VOPs). The basic tools required to encode and decode rectangular I-VOPs and P-VOPs are described in the next section (Section 3.6 of Chapter 3 provides a more detailed 'walk-through' of the encoding and decoding process). The tools in the VLBV Core are based on the H.263 standard and the 'short header' mode enables direct compatibility (at the frame level) between an MPEG-4 Simple Profile CODEC and an H.263 Baseline CODEC.

5.3.2.2 Basic coding tools

I-VOP

A rectangular I-VOP is a frame of video encoded in Intra mode (without prediction from any other coded VOP). The encoding and decoding stages are shown in Figure 5.7.

Table 5.4 Values of *dc_scaler* parameter depending on QP range

Block type	$QP \leq 4$	$5 \leq QP \leq 8$	$9 \leq QP \leq 24$	$25 \leq QP$
Luma	8	$2 \times QP$	$QP + 8$	$(2 \times QP) - 16$
Chroma	8	$(QP + 13)/2$	$(QP + 13)/2$	$QP - 6$

DCT and IDCT: Blocks of luma and chroma samples are transformed using an 8×8 Forward DCT during encoding and an 8×8 Inverse DCT during decoding (see Section 3.4).

Quantisation: The MPEG-4 Visual standard specifies the method of rescaling ('inverse quantising') quantised transform coefficients in a decoder. Rescaling is controlled by a quantiser scale parameter, *QP*, which can take values from 1 to 31 (larger values of *QP* produce a larger quantiser step size and therefore higher compression and distortion). Two methods of rescaling are described in the standard: 'method 2' (basic method) and 'method 1' (more flexible but also more complex). Method 2 inverse quantisation operates as follows. The DC coefficient in an Intra-coded macroblock is rescaled by:

$$DC = DC_Q \cdot dc\_scaler \qquad (5.1)$$

DC_Q is the quantised coefficient, DC is the rescaled coefficient and *dc_scaler* is a parameter defined in the standard. In short header mode (see below), *dc_scaler* is 8 (i.e. all Intra DC coefficients are rescaled by a factor of 8), otherwise *dc_scaler* is calculated according to the value of *QP* (Table 5.4). All other transform coefficients (including AC and Inter DC) are rescaled as follows:

$$
\begin{aligned}
|F| &= QP \cdot (2 \cdot |F_Q| + 1) & &\text{(if } QP \text{ is odd and } F_Q \neq 0) \\
|F| &= QP \cdot (2 \cdot |F_Q| + 1) - 1 & &\text{(if } QP \text{ is even and } F_Q \neq 0) \qquad (5.2) \\
F &= 0 & &\text{(if } F_Q = 0)
\end{aligned}
$$

F_Q is the quantised coefficient and F is the rescaled coefficient. The sign of F is made the same as the sign of F_Q. Forward quantisation is not defined by the standard.

Zig-zag scan: Quantised DCT coefficients are reordered in a zig-zag scan prior to encoding (see Section 3.4).

Last-Run-Level coding: The array of reordered coefficients corresponding to each block is encoded to represent the zero coefficients efficiently. Each nonzero coefficient is encoded as a triplet of (last, run, level), where 'last' indicates whether this is the final nonzero coefficient in the block, 'run' signals the number of preceding zero coefficients and 'level' indicates the coefficient sign and magnitude.

Entropy coding: Header information and (last, run, level) triplets (see Section 3.5) are represented by variable-length codes (VLCs). These codes are similar to Huffman codes and are defined in the standard, based on pre-calculated coefficient probabilities

A coded I-VOP consists of a VOP header, optional video packet headers and coded macroblocks. Each macroblock is coded with a header (defining the macroblock type, identifying which blocks in the macroblock contain coded coefficients, signalling changes in quantisation parameter, etc.) followed by coded coefficients for each 8×8 block.

In the decoder, the sequence of VLCs are decoded to extract the quantised transform coefficients which are re-scaled and transformed by an 8×8 IDCT to reconstruct the decoded I-VOP (Figure 5.7).

P-VOP

A P-VOP is coded with Inter prediction from a previously encoded I- or P-VOP (a reference VOP). The encoding and decoding stages are shown in Figure 5.8.

Motion estimation and compensation: The basic motion compensation scheme is block-based compensation of 16×16 pixel macroblocks (see Chapter 3). The offset between the current macroblock and the compensation region in the reference picture (the motion vector) may have half-pixel resolution. Predicted samples at sub-pixel positions are calculated using bilinear interpolation between samples at integer-pixel positions. The method of motion estimation (choosing the 'best' motion vector) is left to the designer's discretion. The matching region (or prediction) is subtracted from the current macroblock to produce a residual macroblock (Motion-Compensated Prediction, MCP in Figure 5.8).

After motion compensation, the residual data is transformed with the DCT, quantised, reordered, run-level coded and entropy coded. The quantised residual is rescaled and inverse transformed in the encoder in order to reconstruct a local copy of the decoded MB (for further motion compensated prediction). A coded P-VOP consists of VOP header, optional video packet headers and coded macroblocks each containing a header (this time including differentially-encoded motion vectors) and coded residual coefficients for every 8×8 block.

The decoder forms the same motion-compensated prediction based on the received motion vector and its own local copy of the reference VOP. The decoded residual data is added to the prediction to reconstruct a decoded macroblock (Motion-Compensated Reconstruction, MCR in Figure 5.8).

Macroblocks within a P-VOP may be coded in Inter mode (with motion compensated prediction from the reference VOP) or Intra mode (no motion compensated prediction). Inter mode will normally give the best coding efficiency but Intra mode may be useful in regions where there is not a good match in a previous VOP, such as a newly-uncovered region.

Short Header

The 'short header' tool provides compatibility between MPEG-4 Visual and the ITU-T H.263 video coding standard. An I- or P-VOP encoded in 'short header' mode has identical syntax to an I-picture or P-picture coded in the baseline mode of H.263. This means that an MPEG-4 I-VOP or P-VOP should be decodeable by an H.263 decoder and vice versa.

In short header mode, the macroblocks within a VOP are organised in Groups of Blocks (GOBs), each consisting of one or more complete rows of macroblocks. Each GOB may (optionally) start with a resynchronisation marker (a fixed-length binary code that enables a decoder to resynchronise when an error is encountered, see Section 5.3.2.4).

5.3.2.3 Coding Efficiency Tools

The following tools, part of the Simple profile, can improve compression efficiency. They are only used when short header mode is not enabled.

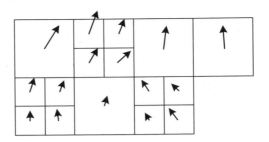

Figure 5.9   One or four vectors per macroblock

Four motion vectors per macroblock

Motion compensation tends to be more effective with smaller block sizes. The default block size for motion compensation is 16×16 samples (luma), 8×8 samples (chroma), resulting in one motion vector per macroblock. This tool gives the encoder the option to choose a smaller motion compensation block size, 8×8 samples (luma) and 4×4 samples (chroma), giving four motion vectors per macroblock. This mode can be more effective at minimising the energy in the motion-compensated residual, particularly in areas of complex motion or near the boundaries of moving objects. There is an increased overhead in sending four motion vectors instead of one, and so the encoder may choose to send one or four motion vectors on a macroblock-by-macroblock basis (Figure 5.9).

Unrestricted Motion Vectors

In some cases, the best match for a macroblock may be a 16×16 region that extends outside the boundaries of the reference VOP. Figure 5.10 shows the lower-left corner of a current VOP (right-hand image) and the previous, reference VOP (left-hand image). The hand holding the bow is moving **into** the picture in the current VOP and so there isn't a good match for the highlighted macroblock inside the reference VOP. In Figure 5.11, the samples in the reference VOP have been extrapolated ('padded') beyond the boundaries of the VOP. A better match for the macroblock is obtained by allowing the motion vector to point into this extrapolated region (the highlighted macroblock in Figure 5.11 is the best match in this case). The Unrestricted Motion Vectors (UMV) tool allows motion vectors to point outside the boundary of the reference VOP. If a sample indicated by the motion vector is outside the reference VOP, the nearest edge sample is used instead. UMV mode can improve motion compensation efficiency, especially when there are objects moving in and out of the picture.

Intra Prediction

Low-frequency transform coefficients of neighbouring intra-coded 8×8 blocks are often correlated. In this mode, the DC coefficient and (optionally) the first row and column of AC coefficients in an Intra-coded 8×8 block are predicted from neighbouring coded blocks. Figure 5.12 shows a macroblock coded in intra mode and the DCT coefficients for each of the four 8×8 luma blocks are shown in Figure 5.13. The DC coefficients (top-left) are clearly

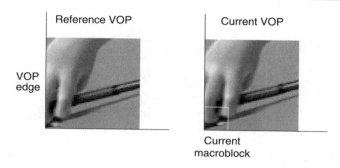

Figure 5.10 Reference VOP and current VOP

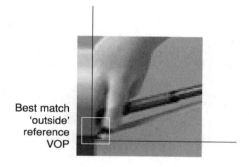

Figure 5.11 Reference VOP extrapolated beyond boundary

Figure 5.12 Macroblock coded in intra mode

similar but it is less obvious whether there is correlation between the first row and column of the AC coefficients in these blocks.

The DC coefficient of the current block (X in Figure 5.14) is predicted from the DC coefficient of the upper (C) or left (A) previously-coded 8×8 block. The rescaled DC coefficient values of blocks A, B and C determine the method of DC prediction. If A, B or C are outside the VOP boundary or the boundary of the current video packet (see later), or if they are not intra-coded, their DC coefficient value is assumed to be equal to

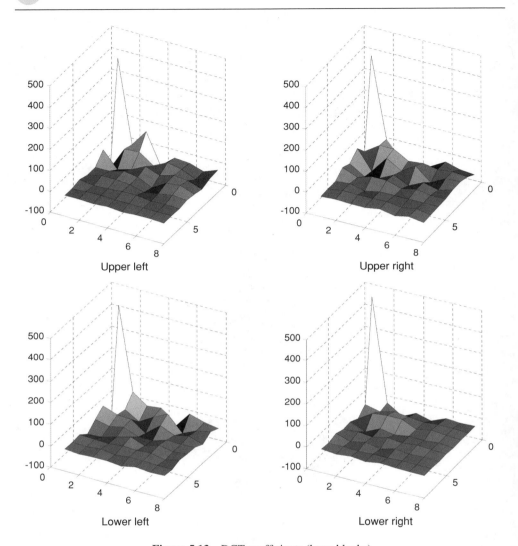

Figure 5.13   DCT coefficients (luma blocks)

1024 (the DC coefficient of a mid-grey block of samples). The *direction* of prediction is determined by:

$$\text{if } |DC_A - DC_B| < |DC_B - DC_C|$$

$$\text{predict from block C}$$

$$\text{else}$$

$$\text{predict from block A}$$

The direction of the smallest DC gradient is chosen as the prediction direction for block X. The prediction, P_{DC}, is formed by dividing the DC coefficient of the chosen neighbouring

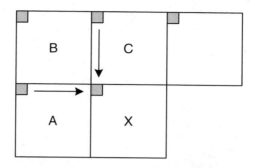

Figure 5.14   Prediction of DC coefficients

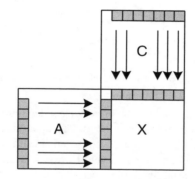

Figure 5.15   Prediction of AC coefficients

block by a scaling factor and P_{DC} is then subtracted from the actual quantised DC coefficient (QDC_X) and the residual ($PQDC_X$) is coded and transmitted.

AC coefficient prediction is carried out in a similar way, with the first row or column of AC coefficients predicted in the direction determined for the DC coefficient (Figure 5.15). For example, if the prediction direction is from block A, the first *column* of AC coefficients in block X is predicted from the first column of block A. If the prediction direction is from block C, the first *row* of AC coefficients in X is predicted from the first row of C. The prediction is scaled depending on the quantiser step sizes of blocks X and A or C.

5.3.2.4 Transmission Efficiency Tools

A transmission error such as a bit error or packet loss may cause a video decoder to lose synchronisation with the sequence of decoded VLCs. This can cause the decoder to decode incorrectly some or all of the information *after* the occurrence of the error and this means that part or all of the decoded VOP will be distorted or completely lost (i.e. the effect of the error spreads spatially through the VOP, 'spatial error propagation'). If subsequent VOPs are predicted from the damaged VOP, the distorted area may be used as a prediction reference, leading to temporal error propagation in subsequent VOPs (Figure 5.16).

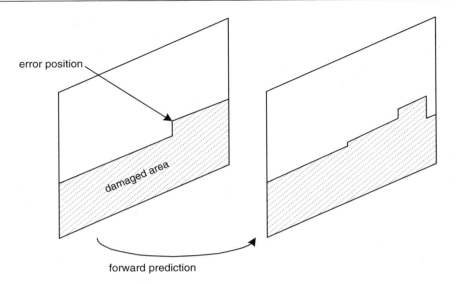

error position

damaged area

forward prediction

Figure 5.16   Spatial and temporal error propagation

When an error occurs, a decoder can resume correct decoding upon reaching a *resynchronisation point*, typically a uniquely-decodeable binary code inserted in the bitstream. When the decoder detects an error (for example because an invalid VLC is decoded), a suitable recovery mechanism is to 'scan' the bitstream until a resynchronisation marker is detected. In short header mode, resynchronisation markers occur at the start of each VOP and (optionally) at the start of each GOB.

The following tools are designed to improve performance during transmission of coded video data and are particularly useful where there is a high probability of network errors [3]. The tools may not be used in short header mode.

Video Packet

A transmitted VOP consists of one or more video packets. A video packet is analogous to a slice in MPEG-1, MPEG-2 or H.264 (see Section 6) and consists of a resynchronisation marker, a header field and a series of coded macroblocks in raster scan order (Figure 5.17). (Confusingly, the MPEG-4 Visual standard occasionally refers to video packets as 'slices'). The resynchronisation marker is followed by a count of the next macroblock number, which enables a decoder to position the first macroblock of the packet correctly. After this comes the quantisation parameter and a flag, HEC (Header Extension Code). If HEC is set to 1, it is followed by a duplicate of the current VOP header, increasing the amount of information that has to be transmitted but enabling a decoder to recover the VOP header if the first VOP header is corrupted by an error.

The video packet tool can assist in error recovery at the decoder in several ways, for example:

1. When an error is detected, the decoder can resynchronise at the start of the next video packet and so the error does not propagate beyond the boundary of the video packet.

Figure 5.17 Video packet structure

2. If used, the HEC field enables a decoder to recover a lost VOP header from elsewhere within the VOP.
3. Predictive coding (such as differential encoding of the quantisation parameter, prediction of motion vectors and intra DC/AC prediction) does not cross the boundary between video packets. This prevents (for example) an error in motion vector data from propagating to another video packet.

Data Partitioning

The *data partitioning* tool enables an encoder to reorganise the coded data within a video packet to reduce the impact of transmission errors. The packet is split into two partitions, the first (immediately after the video packet header) containing coding mode information for each macroblock together with DC coefficients of each block (for Intra macroblocks) or motion vectors (for Inter macroblocks). The remaining data (AC coefficients and DC coefficients of Inter macroblocks) are placed in the second partition following a resynchronisation marker. The information sent in the first partition is considered to be the most important for adequate decoding of the video packet. If the first partition is recovered, it is usually possible for the decoder to make a reasonable attempt at reconstructing the packet, even if the 2nd partition is damaged or lost due to transmission error(s).

Reversible VLCs

An optional set of *Reversible Variable Length Codes* (RVLCs) may be used to encode DCT coefficient data. As the name suggests, these codes can be correctly decoded in both the forward and reverse directions, making it possible for the decoder to minimise the picture area affected by an error.

A decoder first decodes each video packet in the forward direction and, if an error is detected (e.g. because the bitstream syntax is violated), the packet is decoded in the reverse direction from the next resynchronisation marker. Using this approach, the damage caused by an error may be limited to just one macroblock, making it easy to conceal the errored region. Figure 5.18 illustrates the use of error resilient decoding. The figure shows a video packet that uses HEC, data partitioning and RVLCs. An error occurs within the texture data and the decoder scans forward and backward to recover the texture data on either side of the error.

5.3.3 The Advanced Simple Profile

The Simple profile, introduced in the first version of the MPEG-4 Visual standard, rapidly became popular with developers because of its improved efficiency compared with previous standards (such as MPEG-1 and MPEG-2) and the ease of integrating it into existing video applications that use rectangular video frames. The Advanced Simple profile was incorporated

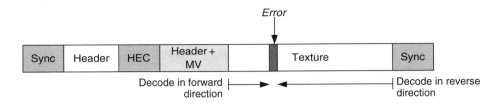

Figure 5.18   Error recovery using RVLCs

into a later version of the standard with added tools to support improved compression efficiency and interlaced video coding. An Advanced Simple Profile CODEC must be capable of decoding Simple objects as well as Advanced Simple objects which may use the following tools in addition to the Simple Profile tools:

- B-VOP (bidirectionally predicted Inter-coded VOP);
- quarter-pixel motion compensation;
- global motion compensation;
- alternate quantiser;
- interlace (tools for coding interlaced video sequences).

B-VOP

The B-VOP uses bidirectional prediction to improve motion compensation efficiency. Each block or macroblock may be predicted using (a) forward prediction from the previous I- or P-VOP, (b) backwards prediction from the next I- or P-VOP or (c) an average of the forward and backward predictions. This mode generally gives better coding efficiency than basic forward prediction; however, the encoder must store multiple frames prior to coding each B-VOP which increases the memory requirements and the encoding delay. Each macroblock in a B-VOP is motion compensated from the previous and/or next I- or P-VOP in one of the following ways (Figure 5.19).

1. Forward prediction: A single MV is transmitted, MV_F, referring to the *previous* I- or P-VOP.
2. Backward prediction: A single MV is transmitted, MV_B, referring to the *future* I- or P-VOP.
3. Bidirectional interpolated prediction: Two MVs are transmitted, MV_F and MV_B, referring to the previous and the future I- or P-VOPs. The motion compensated prediction for the current macroblock is produced by interpolating between the luma and chroma samples in the two reference regions.
4. Bidirectional direct prediction: Motion vectors pointing to the previous and future I- or P-VOPs are derived automatically from the MV of the same macroblock in the *future* I- or P-VOP. A 'delta MV' correcting these automatically-calculated MVs is transmitted.

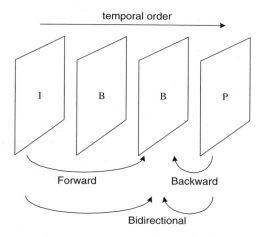

Figure 5.19 Prediction modes for B-VOP

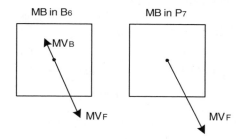

Figure 5.20 Direct mode vectors

Example of direct mode (Figure 5.20)

Previous reference VOP: I_4, display_time = 2
Current B-VOP: B_6, display_time = 6
Future reference VOP: P_7, display_time = 7

MV for same macroblock position in P_7, $MV_7 = (+5, -10)$
TRB = display_time(B_6) − display_time(I_4) = 4
TRD = display_time(P_7) − display_time(I_4) = 5
$MV_D = 0$ (no delta vector)
$MV_F = (TRB/TRD).MV = (+4, -8)$
$MV_B = (TRB-TRD/TRD).MV = (-1, +2)$

Quarter-Pixel Motion Vectors

The Simple Profile supports motion vectors with half-pixel accuracy and this tool supports vectors with quarter-pixel accuracy. The reference VOP samples are interpolated to half-pixel positions and then again to quarter-pixel positions prior to motion estimation and compensation. This increases the complexity of motion estimation, compensation and reconstruction

Table 5.5   Weighting matrix W_w

10	20	20	30	30	30	40	40
20	20	30	30	30	40	40	40
20	30	30	30	40	40	40	40
30	30	30	30	40	40	40	50
30	30	30	40	40	40	50	50
30	40	40	40	40	40	50	50
40	40	40	40	50	50	50	50
40	40	40	50	50	50	50	50

but can provide a gain in coding efficiency compared with half-pixel compensation (see Chapter 3).

Alternate quantiser

An alternative rescaling ('inverse quantisation') method is supported in the Advanced Simple Profile. Intra DC rescaling remains the same (see Section 5.3.2) but other quantised coefficients may be rescaled using an alternative method[1].

Quantised coefficients $F_Q(u, v)$ are rescaled to produce coefficients $F(u, v)$ (where u, $v$are the coordinates of the coefficient) as follows:

$$F = 0 \qquad\qquad\qquad\qquad\qquad \text{if } F_Q = 0$$
$$F = [(2.F_Q(u, v) + k) \cdot W_w(u, v) \cdot QP]/16 \qquad \text{if } F_Q \neq 0 \qquad (5.3)$$

$$k = \begin{cases} 0 & \text{intra blocks} \\ +1 & F_Q(u, v) > 0, \text{ nonintra} \\ -1 & F_{rmQ}(u, v) < 0, \text{ nonintra} \end{cases}$$

where W_W is a matrix of weighting factors, W_0 for intra macroblocks and W_1 for nonintra macroblocks. In Method 2 rescaling (see Section 5.3.2.1), all coefficients (apart from Intra DC) are quantised and rescaled with the same quantiser step size. Method 1 rescaling allows an encoder to vary the step size depending on the position of the coefficient, using the weighting matrix W_W. For example, better subjective performance may be achieved by increasing the step size for high-frequency coefficients and reducing it for low-frequency coefficients. Table 5.5 shows a simple example of a weighting matrix W_W.

Global Motion Compensation

Macroblocks within the same video object may experience similar motion. For example, camera pan will produce apparent linear movement of the entire scene, camera zoom or rotation will produce a more complex apparent motion and macroblocks within a large object may all move in the same direction. Global Motion Compensation (GMC) enables an encoder to transmit a small number of motion (warping) parameters that describe a default 'global' motion for the entire VOP. GMC can provide improved compression efficiency when a significant number of macroblocks in the VOP share the same motion characteristics. The global motion

[1] The MPEG-4 Visual standard describes the default rescaling method as 'Second Inverse Quantisation Method' and the alternative, optional method as 'First Inverse Quantisation Method'. The default ('Second') method is sometimes known as 'H.263 quantisation' and the alternative ('First') method as 'MPEG-4 quantisation'.

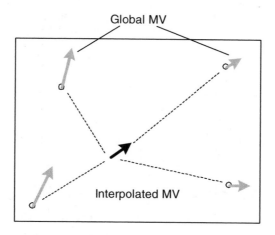

Figure 5.21 VOP, GMVs and interpolated vector

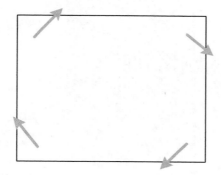

Figure 5.22 GMC (compensating for rotation)

parameters are encoded in the VOP header and the encoder chooses either the default GMC parameters or an individual motion vector for each macroblock.

When the GMC tool is used, the encoder sends up to four *global motion vectors* (GMVs) for each VOP together with the location of each GMV in the VOP. For each pixel position in the VOP, an individual motion vector is calculated by interpolating between the GMVs and the pixel position is motion compensated according to this interpolated vector (Figure 5.21). This mechanism enables compensation for a variety of types of motion including rotation (Figure 5.22), camera zoom (Figure 5.23) and warping as well as translational or linear motion.

The use of GMC is enabled by setting the parameter *sprite_enable* to 'GMC' in a Video Object Layer (VOL) header. VOPs in the VOL may thereafter be coded as S(GMC)-VOPs ('sprite' VOPs with GMC), as an alternative to the 'usual' coding methods (I-VOP, P-VOP or B-VOP). The term 'sprite' is used here because a type of global motion compensation is applied in the older 'sprite coding' mode (part of the Main Profile, see Section 5.4.2.2).

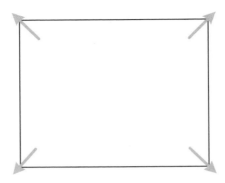

Figure 5.23 GMC (compensating for camera zoom)

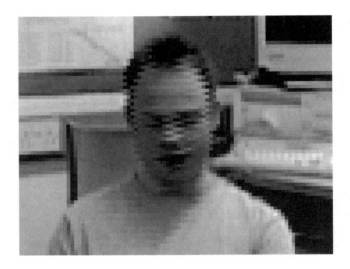

Figure 5.24 Close-up of interlaced VOP

Interlace

Interlaced video consists of two fields per frame (see Chapter 2) sampled at different times (typically at 50 Hz or 60 Hz temporal sampling rate). An interlaced VOP contains alternate lines of samples from two fields. Because the fields are sampled at different times, horizontal movement may reduce correlation between lines of samples (for example, in the moving face in Figure 5.24). The encoder may choose to encode the macroblock in Frame DCT mode, in which each block is transformed as usual, or in Field DCT mode, in which the luminance samples from Field 1 are placed in the top eight lines of the macroblock and the samples from Field 2 in the lower eight lines of the macroblock before calculating the DCT (Figure 5.25). Field DCT mode gives better performance when the two fields are decorrelated.

In Field Motion Compensation mode (similar to 16×8 Motion Compensation Mode in the MPEG-two standard), samples belonging to the two fields in a macroblock are motion

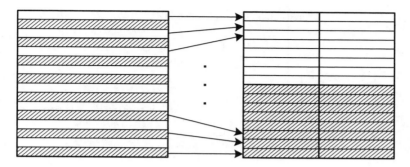

Figure 5.25 Field DCT

compensated separately so that two motion vectors are generated for the macroblock, one for the first field and one for the second. The Direct Mode used in B-VOPs (see above) is modified to deal with macroblocks that have Field Motion Compensated reference blocks. Two forward and two backward motion vectors are generated, one from each field in the forward and backward directions. If the interlaced video tool is used in conjunction with object-based coding (see Section 5.4), the padding process may be applied separately to the two fields of a boundary macroblock.

5.3.4 The Advanced Real Time Simple Profile

Streaming video applications for networks such as the Internet require good compression and error-robust video coding tools that can adapt to changing network conditions. The coding and error resilience tools within Simple Profile are useful for real-time streaming applications and the Advanced Real Time Simple (ARTS) object type adds further tools to improve error resilience and coding flexibility, NEWPRED (multiple prediction references) and Dynamic Resolution Conversion (also known as Reduced Resolution Update). An ARTS Profile CODEC should support Simple and ARTS object types.

NEWPRED

The *NEWPRED* ('new prediction') tool enables an encoder to select a prediction reference VOP from any of a set of previously encoded VOPs for each video packet. A transmission error that is imperfectly concealed will tend to propagate temporally through subsequent predicted VOPs and NEWPRED can be used to limit temporal propagation as follows (Figure 5.26). Upon detecting an error in a decoded VOP (VOP1 in Figure 5.26), the decoder sends a feedback message to the encoder identifying the errored video packet. The encoder chooses a reference VOP *prior* to the errored packet (VOP 0 in this example) for encoding of the following VOP (frame 4). This has the effect of 'cleaning up' the error and halting temporal propagation. Using NEWPRED in this way requires both encoder and decoder to store multiple reconstructed VOPs to use as possible prediction references. Predicting from an older reference VOP (4 VOPs in the past in this example) tends to reduce compression performance because the correlation between VOPs reduces with increasing time.

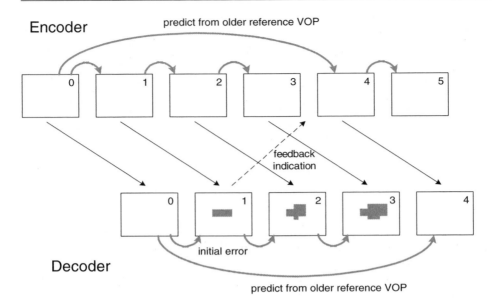

Figure 5.26   NEWPRED error handling

Dynamic Resolution Conversion

Dynamic Resolution Conversion (DRC), otherwise known as Reduced Resolution (RR) mode, enables an encoder to encode a VOP with reduced spatial resolution. This can be a useful tool to prevent sudden increases in coded bitrate due to (for example) increased detail or rapid motion in the scene. Normally, such a change in the scene content would cause the encoder to generate a large number of coded bits, causing problems for a video application transmitting over a limited bitrate channel. Using the DRC tool, a VOP is encoded at half the normal horizontal and vertical resolution. At the decoder, a residual macroblock within a Reduced Resolution VOP is decoded and upsampled (interpolated) so that each 8×8 luma block covers an area of 16×16 samples. The upsampled macroblock (now covering an area of 32×32 luma samples) is motion compensated from a 32×32-sample reference area (the motion vector of the decoded macroblock is scaled up by a factor of 2) (Figure 5.27). The result is that the Reduced Resolution VOP is decoded at half the normal resolution (so that the VOP detail is reduced) with the benefit that the coded VOP requires fewer bits to transmit than a full-resolution VOP.

5.4 CODING ARBITRARY-SHAPED REGIONS

Coding objects of arbitrary shape (see Section 5.2.3) requires a number of extensions to the block-based VLBV core CODEC [4]. Each VOP is coded using motion compensated prediction and DCT-based coding of the residual, with extensions to deal with the special cases introduced by object boundaries. In particular, it is necessary to deal with shape coding, motion compensation and texture coding of arbitrary-shaped video objects.

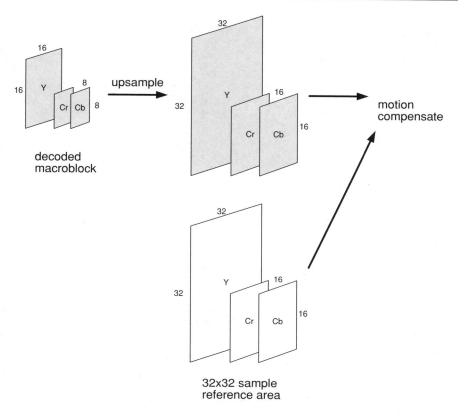

Figure 5.27   Reduced Resolution decoding of a macroblock

Shape coding: The shape of a video object is defined by *Alpha Blocks*, each covering a
16 × 16-pixel area of the video scene. Each Alpha Block may be entirely external to the
video object (in which case nothing needs to be coded), entirely internal to the VO (in
which case the macroblock is encoded as in Simple Profile) or it may cross a boundary
of the VO. In this last case, it is necessary to define the shape of the VO edge within the
Alpha Block. Shape information is defined using the concept of 'transparency', where a
'transparent' pixel is not part of the current VOP, an 'opaque' pixel is part of the VOP
and replaces anything 'underneath' it and a 'semi-transparent' pixel is part of the VOP
and is partly transparent. The shape information may be defined as *binary* (all pixels are
either opaque, 1, or transparent, 0) or *grey scale* (a pixel's transparency is defined by a
number between 0, transparent, and 255, opaque). Binary shape information for a boundary
macroblock is coded as a *binary alpha block* (BAB) using arithmetic coding and grey scale
shape information is coded using motion compensation and DCT-based encoding.

Motion compensation: Each VOP may be encoded as an I-VOP (no motion compensation),
a P-VOP (motion compensated prediction from a past VOP) or a B-VOP (bidirection motion compensated prediction). Nontransparent pixels in a boundary macroblock are motion
compensated from the appropriate reference VOP(s) and the boundary pixels of a reference

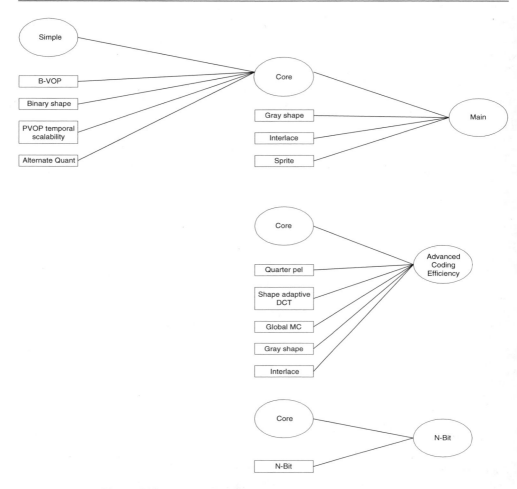

Figure 5.28   Tools and objects for coding arbitrary-shaped regions

VOP are 'padded' to the edges of the motion estimation search area to fill the transparent pixel positions with data.

Texture coding: Motion-compensated residual samples ('texture') in internal blocks are coded using the 8×8 DCT, quantisation and variable length coding described in Section 5.3.2.1. Non-transparent pixels in a boundary block are padded to the edge of the 8×8 block prior to applying the DCT.

Video object coding is supported by the Core and Main profiles, with extra tools in the Advanced Coding Efficiency and N-Bit profiles (Figure 5.28).

5.4.1 The Core Profile

A Core Profile CODEC should be capable of encoding and decoding Simple Video Objects and Core Video Objects. A Core VO may use any of the Simple Profile tools plus the following:

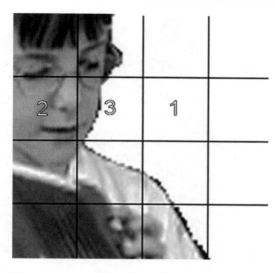

Figure 5.29 VO showing external (1), internal (2) and boundary (3) macroblocks

- B-VOP (described in Section 5.3.3);
- alternate quantiser (described in Section 5.3.3);
- object-based coding (with Binary Shape);
- P-VOP temporal scalability.

Scalable coding, described in detail in Section 5.5, enables a video sequence to be coded and transmitted as two or more separate 'layers' which can be decoded and re-combined. The Core Profile supports temporal scalability using P-VOPs and an encoder using this tool can transmit two coded layers, a base layer (decodeable as a low frame-rate version of the video scene) and a temporal enhancement layer containing only P-VOPs. A decoder can increase the frame rate of the base layer by adding the decoded frames from the enhancement layer.

Probably the most important functionality in the Core Profile is its support for coding of arbitrary-shaped objects, requiring several new tools. Each macroblock position in the picture is classed as (1) *opaque* (fully 'inside' the VOP), (2) *transparent* (not part of the VOP) or (3) on the *boundary* of the VOP (Figure 5.29).

In order to indicate the shape of the VOP to the decoder, *alpha mask* information is sent for every macroblock. In the Core Profile, only *binary* alpha information is permitted and each pixel position in the VOP is defined to be completely opaque or completely transparent. The Core Profile supports coding of binary alpha information and provides tools to deal with the special cases of motion and texture coding within boundary macroblocks.

5.4.1.1 Binary Shape Coding

For each macroblock in the picture, a code *bab_type* is transmitted. This code indicates whether the MB is transparent (not part of the current VOP, therefore no further data should be coded), opaque (internal to the current VOP, therefore motion and texture are coded as usual) or a boundary MB (part of the MB is opaque and part is transparent). Figure 5.30 shows a video object plane and Figure 5.31 is the corresponding *binary mask* indicating which pixels are

Figure 5.30   VOP

Figure 5.31   Binary alpha mask (complete VOP)

part of the VOP (white) and which pixels are outside the VOP (black). For a boundary MB (e.g. Figure 5.32), it is necessary to encode a binary alpha mask to indicate which pixels are transparent and which are opaque (Figure 5.33).

The binary alpha mask (BAB) for each boundary macroblock is coded using Context-based binary Arithmetic Encoding (CAE). A BAB pixel value X is to be encoded, where X is either 0 or 1. First, a *context* is calculated for the current pixel. A *context template* defines a region of n neighbouring pixels that have previously been coded (spatial neighbours for intra-coded BABs, spatial and temporal neighbours for inter-coded BABs). The n values of each

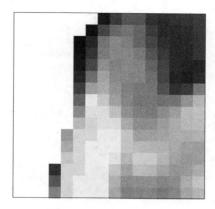

Figure 5.32 Boundary macroblock

Figure 5.33 Binary alpha mask (boundary MB)

BAB pixel in the context form an n-bit word, the context for pixel X. There are 2^n possible contexts and P(0), the probability that X is 0 given a particular context, is stored in the encoder and decoder for each possible n-bit context. Each mask pixel X is coded as follows:

1. Calculate the context for X.
2. Look up the relevant entry in the probability table P(0).
3. Encode X with an arithmetic encoder (see Chapter 3 for an overview of arithmetic coding). The sub-range is $0 \ldots P(0)$ if X is 0 (black), $P(0) \ldots 1.0$ if X is 1 (white).

Intra BAB Encoding

In an intra-coded BAB, the context template for the current mask pixel is formed from 10 spatially neighbouring pixels that have been previously encoded, c_0 to c_9 in Figure 5.34. The context is formed from the 10-bit word $c_9c_8c_7c_6c_5c_4c_3c_2c_1c_0$. Each of the 1024 possible context probabilities is listed in a table in the MPEG-4 Visual standard as an integer in the range 0 to 65535 and the actual probability of zero P(0) is derived by dividing this integer by 65535.

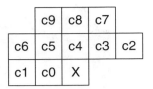

Figure 5.34 Context template for intra BAB

Examples

Context (binary)	Context (decimal)	Description	P(0)
0000000000	0	All context pixels are 0	$65267/65535 = 0.9959$
0000000001	1	c_0 is 1, all others are 0	$16468/65535 = 0.2513$
1111111111	1023	All context pixels are 1	$235/65535 = 0.0036$

The context template (Figure 5.34) extends 2 pixels horizontally and vertically from the position of X. If any of these pixels are undefined (e.g. c_2, c_3 and c_7 may be part of a BAB that has not yet been coded, or some of the pixels may belong to transparent BABs), the undefined pixels are set to the value of the nearest neighbour within the current BAB. Depending on the shape of the binary mask, more efficient coding may be obtained by scanning through the BAB in vertical order (rather than raster order) so that the context template is placed on its 'side'. The choice of scan order for each BAB is signalled in the bitstream.

Inter BAB Encoding

The context template (Figure 5.35) consists of nine pixel positions, four in the current VOP (c_0 to c_3) and five in a reference VOP (c_4 to c_8). The position of the central context pixel in the reference VOP (c_6) may be offset from the position X by an integer sample vector, enabling an inter BAB to be encoded using motion compensation. This 'shape' vector (MV_s) may be chosen independently of any 'texture' motion vector. There are nine context pixels and so a total of $2^9 = 512$ probabilities P(0) are stored by the encoder and decoder.

Examples:

Context (binary)	Context (decimal)	Description	P(0)
000000001	1	c_0 is 1, all others 0	$62970/65535 = 0.9609$
000100000	64	c_6 is 1, all others 0	$23130/65535 = 0.3529$
000100001	65	c_6 and c_0 are 1, all others 0	$7282/65535 = 0.1111$

These examples indicate that the transparency of the current pixel position X is more heavily influenced by c_6 (the same position in the previous motion-compensated BAB) than c_0 (the previous pixel position in raster scan order). As with intra-coding, the scanning order of the current (and previous) BAB may be horizontal or vertical.

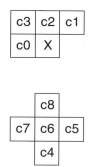

Figure 5.35   Context template for inter BAB

A single vector MV_s is coded for each boundary inter-coded BAB. For P-VOPs, the reference VOP is the previous I- or P-VOP and for a B-VOP, the reference VOP is the temporally 'nearest' I- or P-VOP.

5.4.1.2 Motion Compensated Coding of Arbitrary-shaped VOPs

A P-VOP or B-VOP is predicted from a reference I- or P-VOP using motion compensation. It is possible for a motion vector to point to a reference region that extends outside of the opaque area of the reference VOP, i.e. some of the pixels in the reference region may be 'transparent'. Figure 5.36 illustrates three examples. The left-hand diagram shows a reference VOP (with opaque pixels coloured grey) and the right-hand diagram shows a current VOP consisting of nine macroblocks. MB1 is entirely opaque but its MV points to a region in the reference VOP that contains transparent pixels. MB2 is a boundary MB and the opaque part of its motion-compensated reference region is smaller than the opaque part of MB2. MB3 is also a boundary MB and part of its reference region is located in a completely transparent MB in the reference VOP. In each of these cases, some of the opaque pixels in the current MB are motion-compensated from transparent pixels in the reference VOP. The values of transparent pixels are not defined and so it is necessary to deal with these special cases. This is done by *padding* transparent pixel positions in boundary and transparent macroblocks in the reference VOP.

Padding of Boundary MBs

Transparent pixels in each boundary MB in a reference VOP are extrapolated horizontally and vertically from opaque pixels as shown in Figure 5.37.

1. Opaque pixels at the edge of the BAB (dark grey in Figure 5.37) are extrapolated horizontally to fill transparent pixel positions in the same row. If a row is bordered by opaque pixels at only one side, the value of the nearest opaque pixel is copied to all transparent pixel positions. If a row is bordered on two sides by opaque pixels (e.g. the top row in Figure 5.37(a)), the transparent pixel positions are filled with the mean value of the two neighbouring opaque pixels. The result of horizontal padding is shown in Figure 5.37(b).
2. Opaque pixels (including those 'filled' by the first stage of horizontal padding) are extrapolated vertically to fill the remaining transparent pixel positions. Columns of transparent

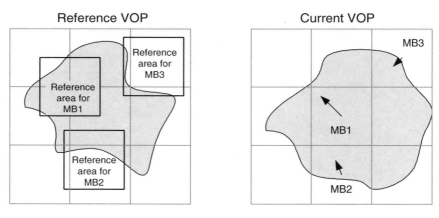

Figure 5.36   Examples of reference areas containing transparent pixels

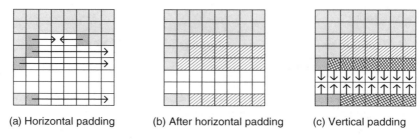

(a) Horizontal padding   (b) After horizontal padding   (c) Vertical padding

Figure 5.37   Horizontal and vertical padding in boundary MB

pixels with one opaque neighbour are filled with the value of that pixel and columns with two opaque neighbours (such as those in Figure 5.37(c)) are filled with the mean value of the opaque pixels at the top and bottom of the column.

Example

Figure 5.38 shows a boundary macroblock from a VOP with transparent pixels plotted in black. The opaque pixels are extrapolated horizontally (step 1) to produce Figure 5.39 (note that five transparent pixel positions have two opaque horizontal neighbours). The result of step 1 is then extrapolated vertically (step 2) to produce Figure 5.40.

Padding of Transparent MBs

Fully transparent MBs must also be filled with padded pixel values because they may fall partly or wholly within a motion-compensated reference region (e.g. the reference region for MB3 in Figure 5.36). Transparent MBs with a single neighbouring boundary MB are filled by horizontal or vertical extrapolation of the boundary pixels of that MB. For example, in Figure 5.41, a transparent MB to the left of the boundary MB shown in Figure 5.38 is padded by filling each pixel position in the transparent MB with the value of the horizontally adjacent

Figure 5.38 Boundary MB

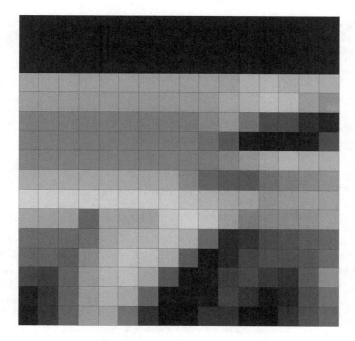

Figure 5.39 Boundary MB after horizontal padding

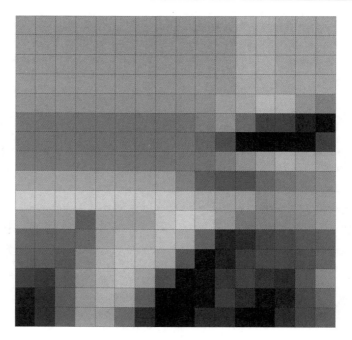

Figure 5.40   Boundary MB after vertical padding

edge pixel. Transparent MBs are always padded *after* all boundary MBs have been fully padded.

If a transparent MB has more than one neighbouring boundary MB, one of its neighbours is chosen for extrapolation according to the following rule. If the left-hand MB is a boundary MB, it is chosen; else if the top MB is a boundary MB, it is chosen; else if the right-hand MB is a boundary MB, it is chosen; else the lower MB is chosen.

Transparent MBs with no nontransparent neighbours are filled with the pixel value 2^{N-1}, where N is the number of bits per pixel. If N is 8 (the usual case), these MBs are filled with the pixel value 128.

5.4.1.3 Texture Coding in Boundary Macroblocks

The texture in an opaque MB (the pixel values in an intra-coded MB or the motion compensated residual in an inter-coded MB) is coded by the usual process of 8×8 DCT, quantisation, run-level encoding and entropy encoding (see Section 5.3.2). A boundary MB consists partly of texture pixels (inside the boundary) and partly of undefined, transparent pixels (outside the boundary). In a core profile object, each 8×8 texture block within a boundary MB is coded using an 8×8 DCT followed by quantisation, run-level coding and entropy coding as usual (see Section 7.2 for an example). (The Shape-Adaptive DCT, part of the Advanced Coding Efficiency Profile and described in Section 5.4.3, provides a more efficient method of coding boundary texture.)

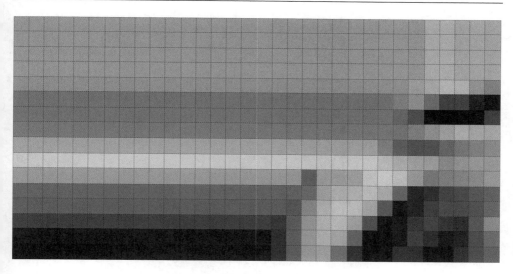

Figure 5.41 Padding of transparent MB from horizontal neighbour

5.4.2 The Main Profile

A Main Profile CODEC supports Simple and Core objects plus Scalable Texture objects (see Section 5.6.1) and Main objects. The Main object adds the following tools:

- interlace (described in Section 5.3.3);
- object-based coding with grey ('alpha plane') shape;
- Sprite coding.

In the Core Profile, object shape is specified by a binary alpha mask such that each pixel position is marked as 'opaque' or 'transparent'. The Main Profile adds support for grey shape masks, in which each pixel position can take varying levels of transparency from fully transparent to fully opaque. This is similar to the concept of Alpha Planes used in computer graphics and allows the overlay of multiple semi-transparent objects in a reconstructed (rendered) scene.

Sprite coding is designed to support efficient coding of background objects. In many video scenes, the background does not change significantly and those changes that do occur are often due to camera movement. A 'sprite' is a video object (such as the scene background) that is fully or partly transmitted at the start of a scene and then may change in certain limited ways during the scene.

5.4.2.1 Grey Shape Coding

Binary shape coding (described in Section 5.4.1.1) has certain drawbacks in the representation of video scenes made up of multiple objects. Objects or regions in a 'natural' video scene may be translucent (partially transparent) but binary shape coding only supports completely transparent ('invisible') or completely opaque regions. It is often difficult or impossible to segment video objects neatly (since object boundaries may not exactly correspond with pixel positions), especially when segmentation is carried out automatically or semi-automatically.

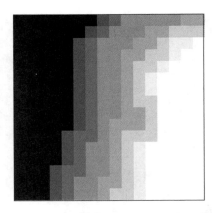

Figure 5.42   Grey-scale alpha mask for boundary MB

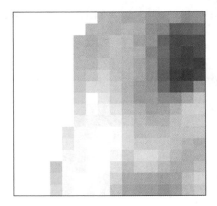

Figure 5.43   Boundary MB with grey-scale transparency

For example, the edge of the VOP shown in Figure 5.30 is not entirely 'clean' and this may lead to unwanted artefacts around the VOP edge when it is rendered with other VOs.

Grey shape coding gives more flexible control of object transparency. A grey-scale alpha plane is coded for each macroblock, in which each pixel position has a mask value between 0 and 255, where 0 indicates that the pixel position is fully transparent, 255 indicates that it is fully opaque and other values specify an intermediate level of transparency. An example of a grey-scale mask for a boundary MB is shown in Figure 5.42. The transparency ranges from fully transparent (black mask pixels) to opaque (white mask pixels). The rendered MB is shown in Figure 5.43 and the edge of the object now 'fades out' (compare this figure with Figure 5.32). Figure 5.44 is a scene constructed of a background VO (rectangular) and two foreground VOs. The foreground VOs are identical except for their transparency, the left-hand VO uses a binary alpha mask and the right-hand VO has a grey alpha mask which helps the right-hand VO to blend more smoothly with the background. Other uses of grey shape coding include representing translucent objects, or deliberately altering objects to make them semi-transparent (e.g. the synthetic scene in Figure 5.45).

Figure 5.44 Video scene with binary-alpha object (left) and grey-alpha object (right)

Figure 5.45 Video scene with semi-transparent object

Grey scale alpha masks are coded using two components, a *binary support mask* that indicates which pixels are fully transparent (external to the VO) and which pixels are semi- or fully-opaque (internal to the VO), and a *grey scale alpha plane*. Figure 5.33 is the binary support mask for the grey-scale alpha mask of Figure 5.42. The binary support mask is coded in the same way as a BAB (see Section 5.4.1.1). The grey scale alpha plane (indicating the level of transparency of the internal pixels) is coded separately in the same way as object texture (i.e. each 8 × 8 block within the alpha plane is transformed using the DCT, quantised,

Figure 5.46 Sequence of frames

reordered, run-level and entropy coded). The decoder reconstructs the grey scale alpha plane (which may not be identical to the original alpha plane due to quantisation distortion) and the binary support mask. If the binary support mask indicates that a pixel is outside the VO, the corresponding grey scale alpha plane value is set to zero. In this way, the object boundary is accurately preserved (since the binary support mask is losslessly encoded) whilst the decoded grey scale alpha plane (and hence the transparency information) may not be identical to the original.

The increased flexibility provided by grey scale alpha shape coding is achieved at a cost of reduced compression efficiency. Binary shape coding requires the transmission of BABs for each boundary MB and in addition, grey scale shape coding requires the transmission of grey scale alpha plane data for every MB that is semi-transparent.

5.4.2.2 Static Sprite Coding

Three frames from a video sequence are shown in Figure 5.46. Clearly, the background does not change during the sequence (the camera position is fixed). The background (Figure 5.47) may be coded as a *static sprite*. A static sprite is treated as a texture image that may move or warp in certain limited ways, in order to compensate for camera changes such as pan, tilt, rotation and zooming. In a typical scenario, a sprite may be much larger than the visible area of the scene. As the camera 'viewpoint' changes, the encoder transmits parameters indicating how the sprite should be moved and warped to recreate the appropriate visible area in the decoded scene. Figure 5.48 shows a background sprite (the large region) and the area viewed by the camera at three different points in time during a video sequence. As the sequence progresses, the sprite is moved, rotated and warped so that the visible area changes appropriately. A sprite may have arbitrary shape (Figure 5.48) or may be rectangular.

The use of static sprite coding is indicated by setting sprite_enable to 'Static' in a VOL header, after which static sprite coding is used throughout the VOP. The first VOP in a static sprite VOL is an I-VOP and this is followed by a series of S-VOPs (Static Sprite VOPs). Note that a Static Sprite S-VOP is coded differently from a Global Motion Compensation S(GMC)-VOP (described in Section 5.3.3).There are two methods of transmitting and manipulating sprites, a 'basic' sprite (sent in its entirety at the start of a sequence) and a 'low-latency' sprite (updated piece by piece during the sequence).

Figure 5.47   Background sprite

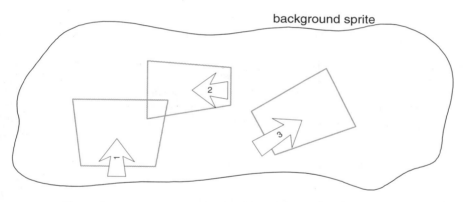

Figure 5.48   Background sprite and three different camera viewpoints

Basic Sprite

The first VOP (I-VOP) contains the entire sprite, encoded in the same way as a 'normal' I-VOP. The sprite may be larger than the visible display size (to accommodate camera movements during the sequence). At the decoder, the sprite is placed in a Sprite Buffer and is not immediately displayed. All further VOPs in the VOL are S-VOPs. An S-VOP contains up to four warping parameters that are used to move and (optionally) warp the contents of the Sprite Buffer in order to produce the desired background display. The number of warping parameters per S-VOP (up to four) is chosen in the VOL header and determines the flexibility of the Sprite Buffer transformation. A single parameter per S-VOP enables linear translation (i.e. a single motion vector for the entire sprite), two or three parameters enable affine transformation of the sprite (e.g. rotation, shear) and four parameters enable a perspective transform.

Low-latency sprite

Transmitting an entire sprite in Basic Sprite mode at the start of a VOL may introduce significant latency because the sprite may be much larger than an individual displayed VOP. The Low-Latency Sprite mode enables an encoder to send initially a minimal size and/or low-quality version of the sprite and then update it during transmission of the VOL. The first I-VOP contains part or all of the sprite (optionally encoded at a reduced quality to save bandwidth) together with the height and width of the entire sprite.

Each subsequent S-VOP may contain warping parameters (as in the Basic Sprite mode) and one or more sprite 'pieces'. A sprite 'piece' covers a rectangular area of the sprite and contains macroblock data that (a) constructs part of the sprite that has not previously been decoded ('static-sprite-object' piece) or (b) improves the quality of part of the sprite that has been previously decoded ('static-sprite-update' piece). Macroblocks in a 'static-sprite-object' piece are encoded as intra macroblocks (including shape information if the sprite is not rectangular). Macroblocks in a 'static-sprite-update' piece are encoded as inter macroblocks using forward prediction from the previous contents of the sprite buffer (but without motion vectors or shape information).

> ### Example
>
> The sprite shown in Figure 5.47 is to be transmitted in low-latency mode. The initial I-VOP contains a low-quality version of part of the sprite and Figure 5.49 shows the contents of the sprite buffer after decoding the I-VOP. An S-VOP contains a new piece of the sprite, encoded in high-quality mode (Figure 5.50) and this extends the contents of the sprite buffer (Figure 5.51). A further S-VOP contains a residual piece (Figure 5.52) that improves the quality of the top-left part of the current sprite buffer. After adding the decoded residual, the sprite buffer contents are as shown Figure 5.53. Finally, four warping points are transmitted in a further S-VOP to produce a change of rotation and perspective (Figure 5.54).

5.4.3 The Advanced Coding Efficiency Profile

The ACE profile is a superset of the Core profile that supports coding of grey-alpha video objects with high compression efficiency. In addition to Simple and Core objects, it includes the ACE object which adds the following tools:

- quarter-pel motion compensation (Section 5.3.3);
- GMC (Section 5.3.3);
- interlace (Section 5.3.3);
- grey shape coding (Section 5.4.2);
- shape-adaptive DCT.

The *Shape-Adaptive DCT* (SA-DCT) is based on pre-defined sets of one-dimensional DCT basis functions and allows an arbitrary region of a block to be efficiently transformed and compressed. The SA-DCT is only applicable to 8×8 blocks within a boundary BAB that

Figure 5.49 Low-latency sprite: decoded I-VOP

Figure 5.50 Low-latency sprite: static-sprite-object piece

Figure 5.51 Low-latency sprite: buffer contents (1)

Figure 5.52   Low-latency sprite: static-sprite-update piece

Figure 5.53   Low-latency sprite: buffer contents (2)

Figure 5.54   Low-latency sprite: buffer contents (3)

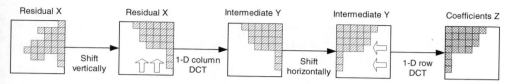

Figure 5.55 Shape-adaptive DCT

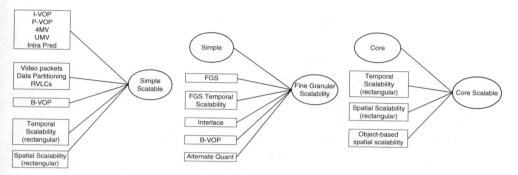

Figure 5.56 Tools and objects for scalable coding

contain one or more transparent pixels. The Forward SA-DCT consists of the following steps (Figure 5.55):

1. Shift opaque residual values X to the top of the 8×8 block.
2. Apply a 1D DCT to each column (the number of points in the transform matches the number of opaque values in each column).
3. Shift the resulting intermediate coefficients Y to the left of the block.
4. Apply a 1D DCT to each row (matched to the number of values in each row).

The final coefficients (Z) are quantised, zigzag scanned and encoded. The decoder reverses the process (making use of the shape information decoded from the BAB) to reconstruct the 8×8 block of samples. The SA-DCT is more complex than the normal 8×8 DCT but can improve coding efficiency for boundary MBs.

5.4.4 The N-bit Profile

The N-bit profile contains Simple and Core objects plus the *N-bit* tool. This supports coding of luminance and chrominance data containing between four and twelve bits per sample (instead of the usual restriction to eight bits per sample). Possible applications of the N-bit profile include video coding for displays with low colour depth (where the limited display capability means that less than eight bits are required to represent each sample) or for high-quality display applications (where the display has a colour depth of more than eight bits per sample and high coded fidelity is desired).

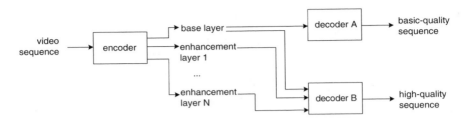

Figure 5.57   Scalable coding: general concept

5.5 SCALABLE VIDEO CODING

Scalable encoding of video data enables a decoder to decode selectively only part of the coded bitstream. The coded stream is arranged in a number of *layers*, including a 'base' layer and one or more 'enhancement' layers (Figure 5.57). In this figure, decoder A receives only the base layer and can decode a 'basic' quality version of the video scene, whereas decoder B receives all layers and decodes a high quality version of the scene. This has a number of applications, for example, a low-complexity decoder may only be capable of decoding the base layer; a low-rate bitstream may be extracted for transmission over a network segment with limited capacity; and an error-sensitive base layer may be transmitted with higher priority than enhancement layers.

MPEG-4 Visual supports a number of scalable coding modes. *Spatial scalability* enables a (rectangular) VOP to be coded at a hierarchy of spatial resolutions. Decoding the base layer produces a low-resolution version of the VOP and decoding successive enhancement layers produces a progressively higher-resolution image. *Temporal scalability* provides a low frame-rate base layer and enhancement layer(s) that build up to a higher frame rate. The standard also supports *quality scalability*, in which the enhancement layers improve the visual quality of the VOP and *complexity scalability*, in which the successive layers are progressively more complex to decode. *Fine Grain Scalability* (FGS) enables the quality of the sequence to be increased in small steps. An application for FGS is streaming video across a network connection, in which it may be useful to scale the coded video stream to match the available bit rate as closely as possible.

5.5.1 Spatial Scalability

The base layer contains a reduced-resolution version of each coded frame. Decoding the base layer alone produces a low-resolution output sequence and decoding the base layer with enhancement layer(s) produces a higher-resolution output. The following steps are required to *encode* a video sequence into two spatial layers:

1. Subsample each input video frame (Figure 5.58) (or video object) horizontally and vertically (Figure 5.59).
2. Encode the reduced-resolution frame to form the base layer.
3. Decode the base layer and up-sample to the original resolution to form a prediction frame (Figure 5.60).
4. Subtract the full-resolution frame from this prediction frame (Figure 5.61).
5. Encode the difference (residual) to form the enhancement layer.

Figure 5.58   Original video frame

Figure 5.59   Sub-sampled frame to be encoded as base layer

Figure 5.60   Base layer frame (decoded and upsampled)

Figure 5.61   Residual to be encoded as enhancement layer

A single-layer decoder decodes only the base layer to produce a reduced-resolution output sequence. A two-layer decoder can reconstruct a full-resolution sequence as follows:

1. Decode the base layer and up-sample to the original resolution.
2. Decode the enhancement layer.
3. Add the decoded residual from the enhancement layer to the decoded base layer to form the output frame.

An **I-VOP** in an enhancement layer is encoded without any spatial prediction, i.e. as a complete frame or object at the enhancement resolution. In an enhancement layer *P-VOP*, the decoded, up-sampled base layer VOP (at the same position in time) is used as a prediction *without* any motion compensation. The difference between this prediction and the input frame is encoded using the texture coding tools, i.e. no motion vectors are transmitted for an enhancement P-VOP. An enhancement layer **B-VOP** is predicted from two directions. The backward prediction is formed by the decoded, up-sampled base layer VOP (at the same position in time), without any motion compensation (and hence without any MVs). The forward prediction is formed by the previous VOP in the enhancement layer (even if this is itself a B-VOP), with motion-compensated prediction (and hence MVs).

If the VOP has arbitrary (binary) shape, a base layer and enhancement layer BAB is required for each MB. The base layer BAB is encoded as usual, based on the shape and size of the base layer object. A BAB in a P-VOP enhancement layer is coded using prediction from an up-sampled version of the base layer BAB. A BAB in a B-VOP enhancement layer may be coded in the same way, or using forward prediction from the previous enhancement VOP (as described in Section 5.4.1.1).

5.5.2 Temporal Scalability

The base layer of a temporal scalable sequence is encoded at a low video frame rate and a temporal enhancement layer consists of I-, P- and/or B-VOPs that can be decoded together with the base layer to provide an increased video frame rate. Enhancement layer VOPs are predicted using motion-compensated prediction according to the following rules.

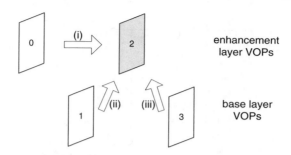

Figure 5.62 Temporal enhancement P-VOP prediction options

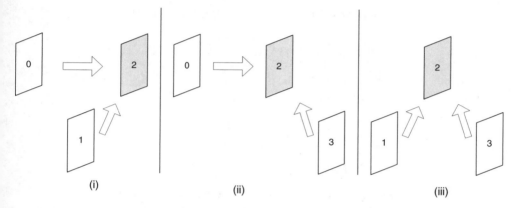

Figure 5.63 Temporal enhancement B-VOP prediction options

An enhancement *I-VOP* is encoded without any prediction. An enhancement *P-VOP* is predicted from (i) the previous enhancement VOP, (ii) the previous base layer VOP or (iii) the next base layer VOP (Figure 5.62). An enhancement *B-VOP* is predicted from (i) the previous enhancement and previous base layer VOPs, (ii) the previous enhancement and next base layer VOPs or (iii) the previous and next base layer VOPs (Figure 5.63).

5.5.3 Fine Granular Scalability

Fine Granular Scalability (FGS) [5] is a method of encoding a sequence as a base layer and enhancement layer. The enhancement layer can be truncated during or after encoding (reducing the bitrate and the decoded quality) to give highly flexible control over the transmitted bitrate. FGS may be useful for video streaming applications, in which the available transmission bandwidth may not be known in advance. In a typical scenario, a sequence is coded as a base layer and a high-quality enhancement layer. Upon receiving a request to send the sequence at a particular bitrate, the streaming server transmits the base layer and a truncated version of the enhancement layer. The amount of truncation is chosen to match the available transmission bitrate, hence maximising the quality of the decoded sequence without the need to re-encode the video clip.

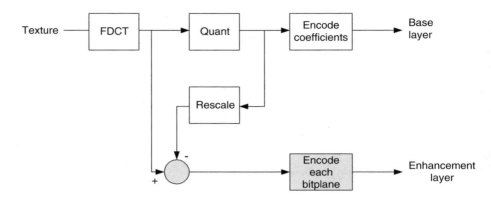

Figure 5.64   FGS encoder block diagram (simplified)

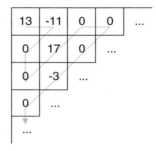

Figure 5.65   Block of residual coefficients (top-left corner)

Encoding

Figure 5.64 shows a simplified block diagram of an FGS encoder (motion compensation is not shown). In the Base Layer, the texture (after motion compensation) is transformed with the forward DCT, quantised and encoded. The quantised coefficients are re-scaled ('inverse quantised') and these re-scaled coefficients are subtracted from the unquantised DCT coefficients to give a set of difference coefficients. The difference coefficients for each block are encoded as a series of *bitplanes*. First, the residual coefficients are reordered using a zigzag scan. The highest-order bits of each coefficient (zeros or ones) are encoded first (the MS bitplane) followed by the next highest-order bits and so on until the LS bits have been encoded.

Example

A block of residual coefficients is shown in Figure 5.65 (coefficients not shown are zero). The coefficients are reordered in a zigzag scan to produce the following list:

$$+13, -11, 0, 0, +17, 0, 0, 0, -3, 0, 0 \ldots$$

The bitplanes corresponding to the magnitude of each residual coefficient are shown in Table 5.6. In this case, the highest plane containing nonzero bits is plane 4 (because the highest magnitude is 17).

Table 5.6 Residual coefficient bitplanes (magnitude)

Value	+13	−11	0	0	+17	0	0	0	−3	0...
Plane 4 (MSB)	0	0	0	0	1	0	0	0	0	0...
Plane 3	1	1	0	0	0	0	0	0	0	0...
Plane 2	1	1	0	0	0	0	0	0	0	0...
Plane 1	0	0	0	0	0	0	0	0	1	0...
Plane 0 (LSB)	1	1	0	0	1	0	0	0	1	0...

Table 5.7 Encoded values

Plane	Encoded values
4	(4, EOP) (+)
3	(0) (+) (0, EOP) (−)
2	(0, EOP)
1	(1) (6, EOP) (−)
0	(0) (0) (2) (3, EOP)

Each bitplane contains a series of zeros and ones. The ones are encoded as (run, EOP) where 'EOP' indicates 'end of bitplane' and each (run, EOP) pair is transmitted as a variable-length code. Whenever the MS bit of a coefficient is encoded, it is immediately followed in the bitstream by a sign bit. Table 5.7 lists the encoded values for each bitplane. Bitplane 4 contains four zeros, followed by a 1. This is the last nonzero bit and so is encoded as (4, EOP). This also the MS bit of the coefficient '+17' and so the sign of this coefficient is encoded.

This example illustrates the processing of one block. The encoding procedure for a complete frame is as follows:

1. Find the highest bit position of any difference coefficient in the frame (the MSB).
2. Encode each bitplane as described above, starting with the plane containing the MSB.

Each complete encoded bitplane is preceded by a start code, making it straightforward to truncate the bitstream by sending only a limited number of encoded bitplanes.

Decoding

The decoder decodes the base layer and enhancement layer (which may be truncated). The difference coefficients are reconstructed from the decoded bitplanes, added to the base layer coefficients and inverse transformed to produce the decoded enhancement sequence (Figure 5.66).

If the enhancement layer has been truncated, then the accuracy of the difference coefficients is reduced. For example, assume that the enhancement layer described in the above example is truncated after bitplane 3. The MS bits (and the sign) of the first three nonzero coefficients are decoded (Table 5.8); if the remaining (undecoded) bitplanes are filled with

Table 5.8   Decoded values (truncated after plane 3)

Plane 4 (MSB)	0	0	0	0	1	0	0	0	0	0...
Plane 3		1	1	0	0	0	0	0	0	0	0...
Plane 2		0	0	0	0	0	0	0	0	0	0...
Plane 1		0	0	0	0	0	0	0	0	0	0...
Plane 0 (LSB)		0	0	0	0	0	0	0	0	0	0...
Decoded value	+8	−8	0	0	+16	0	0	0	0	0...

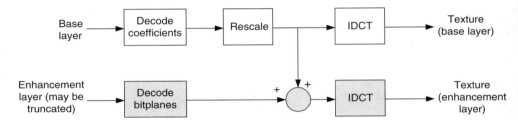

Figure 5.66   FGS decoder block diagram (simplified)

zeros then the list of output values becomes:

$$+8, -8, 0, 0, +16, 0 \ldots.$$

Optional enhancements to FGS coding include selective enhancement (in which bit planes of selected MBs are bit-shifted up prior to encoding, in order to give them a higher priority and a higher probability of being included in a truncated bitstream) and frequency weighting (in which visually-significant low frequency DCT coefficients are shifted up prior to encoding, again in order to give them higher priority in a truncated bitstream).

5.5.4 The Simple Scalable Profile

The Simple Scalable profile supports Simple and Simple Scalable objects. The Simple Scalable object contains the following tools:

- I-VOP, P-VOP, 4MV, unrestricted MV and Intra Prediction;
- Video packets, Data Partitioning and Reversible VLCs;
- B-VOP;
- Rectangular Temporal Scalability (1 enhancement layer) (Section 5.5.2);
- Rectangular Spatial Scalability (1 enhancement layer) (Section 5.5.1).

The last two tools support scalable coding of rectangular VOs.

5.5.5 The Core Scalable Profile

The Core Scalable profile includes Simple, Simple Scalable and Core objects, plus the Core Scalable object which features the following tools, in each case with up to two enhancement layers per object:

- Rectangular Temporal Scalability (Section 5.5.2);
- Rectangular Spatial Scalability (Section 5.5.1);
- Object-based Spatial Scalability (Section 5.5.1).

5.5.6 The Fine Granular Scalability Profile

The FGS profile includes Simple and Advanced Simple objects plus the FGS object which includes these tools:

- B-VOP, Interlace and Alternate Quantiser tools;
- FGS Spatial Scalability;
- FGS Temporal Scalability.

FGS 'Spatial Scalability' uses the encoding and decoding techniques described in Section 5.5.3 to encode each frame as a base layer and an FGS enhancement layer. FGS 'Temporal Scalability' combines FGS (Section 5.5.3) with temporal scalability (Section 5.5.2). An enhancement-layer frame is encoded using forward or bidirectional prediction from *base layer* frame(s) only. The DCT coefficients of the enhancement-layer frame are encoded in bitplanes using the FGS technique.

5.6 TEXTURE CODING

The applications targeted by the developers of MPEG4 include scenarios where it is necessary to transmit still texture (i.e. still images). Whilst block transforms such as the DCT are widely considered to be the best practical solution for motion-compensated video coding, the Discrete Wavelet Transform (DWT) is particularly effective for coding still images (see Chapter 3) and MPEG-4 Visual uses the DWT as the basis for tools to compress still texture. Applications include the coding of rectangular texture objects (such as complete image frames), coding of arbitrary-shaped texture regions and coding of texture to be mapped onto animated 2D or 3D meshes (see Section 5.8).

The basic structure of a still texture encoder is shown in Figure 5.68. A 2D DWT is applied to the texture object, producing a DC component (low-frequency subband) and a number of AC (high-frequency) subbands (see Chapter 3). The DC subband is quantised, predictively encoded (using a form of DPCM) and entropy encoded using an arithmetic encoder. The AC subbands are quantised and reordered ('scanned'), zero-tree encoded and entropy encoded.

Discrete Wavelet Transform

The DWT adopted for MPEG-4 Still Texture coding is the Daubechies (9,3)-tap biorthogonal filter [6]. This is essentially a matched pair of filters, one low pass (with three filter coefficients or 'taps') and one high pass (with nine filter taps).

Quantisation

The DC subband is quantised using a scalar quantiser (see Chapter 3). The AC subbands may be quantised in one of three ways:

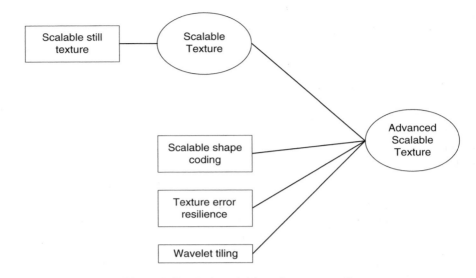

Figure 5.67   Tools and objects for texture coding

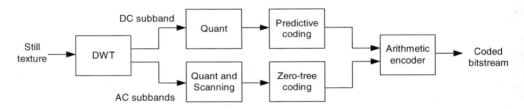

Figure 5.68   Wavelet still texture encoder block diagram

1. Scalar quantisation using a single quantiser ('mode 1'), prior to reordering and zero-tree encoding.
2. 'Bilevel' quantisation ('mode 3') after reordering. The reordered coefficients are coded one bitplane at a time (see Section 5.5.3 for a discussion of bitplanes) using zero-tree encoding. The coded bitstream can be truncated at any point to provide highly scalable decoding (in a similar way to FGS, see previous section).
3. 'Multilevel' quantisation ('mode 2') prior to reordering and zero-tree encoding. A series of quantisers are applied, from coarse to fine, with the output of each quantiser forming a series of layers (a type of scalable coding).

Reordering
The coefficients of the AC subbands are scanned or reordered in one of two ways:

1. Tree-order. A 'parent' coefficient in the lowest subband is coded first, followed by its 'child' coefficients in the next higher subband, and so on. This enables the EZW coding (see below) to exploit the correlation between parent and child coefficients. The first three trees to be coded in a set of coefficients are shown in Figure 5.69.

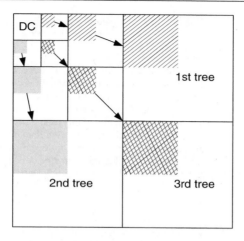

Figure 5.69   Tree-order scanning

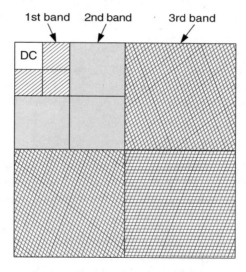

Figure 5.70   Band-by-band scanning

2. Band-by-band order. All the coefficients in the first AC subband are coded, followed by all
   the coefficients in the next subband, and so on (Figure 5.70). This scanning method tends to
   reduce coding efficiency but has the advantage that it supports a form of spatial scalability
   since a decoder can extract a reduced-resolution image by decoding a limited number of
   subbands.

DC Subband Coding

The coefficients in the DC subband are encoded using DPCM. Each coefficient is spatially
predicted from neighbouring, previously-encoded coefficients.

Table 5.9 Zero-tree coding symbols

Symbol	Meaning
ZeroTree Root (ZTR)	The current coefficient and all subsequent coefficients in the tree (or band) are zero. No further data is coded for this tree (or band).
Value + ZeroTree Root (VZTR)	The current coefficient is nonzero but all subsequent coefficients are zero. No further data is coded for this tree/band.
Value (VAL)	The current coefficient is nonzero and one or more subsequent coefficients are nonzero. Further data must be coded.
Isolated Zero (IZ)	The current coefficient is zero but one or more subsequent coefficients are nonzero. Further data must be coded.

AC Subband Coding

Coding of coefficients in the AC subbands is based on EZW (Embedded Zerotree Wavelet coding). The coefficients of each tree (or each subband if band-by-band scanning is used) are encoded starting with the first coefficient (the 'root' of the tree if tree-order scanning is used) and each coefficient is coded as one of the four symbols listed in Table 5.9.

Entropy Coding

The symbols produced by the DC and AC subband encoding processes are entropy coded using a context-based arithmetic encoder. Arithmetic coding is described in Chapter 3 and the principle of context-based arithmetic coding is discussed in Section 5.4.1 and Chapter 6.

5.6.1 The Scalable Texture Profile

The Scalable Texture Profile contains just one object which in turn contains one tool, Scalable Texture. This tool supports the coding process described in the preceding section, for rectangular video objects only. By selecting the scanning mode and quantiser method it is possible to achieve several types of scalable coding.

(a) Single quantiser, tree-ordered scanning: no scalability.
(b) Band-by-band scanning: spatial scalability (by decoding a subset of the bands).
(c) Bilevel quantiser: bitplane-based scalability, similar to FGS.
(d) Multilevel quantiser: 'quality' scalability, with one layer per quantiser.

5.6.2 The Advanced Scalable Texture Profile

The Advanced Scalable Texture profile contains the Advanced Scalable Texture object which adds extra tools to Scalable Texture. *Wavelet tiling* enables an image to be divided into several nonoverlapping sub-images or 'tiles', each coded using the wavelet texture coding process described above. This tool is particularly useful for CODECs with limited memory, since the wavelet transform and other processing steps can be applied to a subset of the image at a time. The *shape coding* tool adds object-based capabilities to the still texture coding process by adapting the DWT to deal with arbitrary-shaped texture objects. Using the *error*

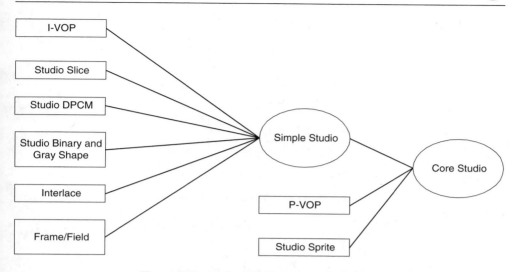

Figure 5.71 Tools and objects for studio coding

resilience tool, the coded texture is partitioned into packets ('texture packets'). The bitstream is processed in Texture Units (TUs), each containing a DC subband, a complete coded tree structure (tree-order scanning) or a complete coded subband (band-by-band scanning). A texture packet contains one or more coded TUs. This packetising approach helps to minimise the effect of a transmission error by localising it to one decoded TU.

5.7 CODING STUDIO-QUALITY VIDEO

Before broadcasting digital video to the consumer it is necessary to code (or transcode) the material into a compressed format. In order to maximise the quality of the video delivered to the consumer it is important to maintain high quality during capture, editing and distribution between studios. The Simple Studio and Core Studio profiles of MPEG-4 Visual are designed to support coding of video at a very high quality for the studio environment. Important considerations include maintaining high fidelity (with near-lossless or lossless coding), support for 4:4:4 and 4:2:2 colour depths and ease of transcoding (conversion) to/from legacy formats such as MPEG-2.

5.7.1 The Simple Studio Profile

The Simple Studio object is intended for use in the capture, storage and editing of high quality video. It supports only I-VOPs (i.e. no temporal prediction) and the coding process is modified in a number of ways.

Source format: The Simple Studio profile supports coding of video sampled in 4:2:0, 4:2:2 and 4:4:4 YCbCr formats (see Chapter 2 for details of these sampling modes) with progressive

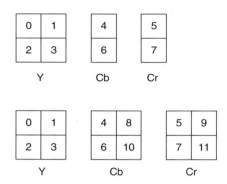

4:4:4 macroblock structure (12 blocks)

Figure 5.72   Modified macroblock structures (4:2:2 and 4:4:4 video)

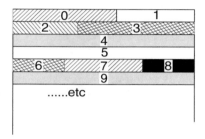

Figure 5.73   Example slice structure

or interlaced scanning. The modified macroblock structures for 4:2:2 and 4:4:4 video are shown in Figure 5.72.

Transform and quantisation: The precision of the DCT and IDCT are extended by three fractional bits. Together with modifications to the forward and inverse quantisation processes, this enables fully lossless DCT-based encoding and decoding. In some cases, lossless DCT coding of intra data may result in a coded frame that is *larger* than the original and for this reason the encoder may optionally use DPCM to code the frame data instead of the DCT (see Chapter 3).

Shape coding: Binary shape information is coded using PCM rather than arithmetic coding (in order to simplify the encoding and decoding process). Alpha (grey) shape may be coded with an extended resolution of up to 12 bits.

Slices: Coded data are arranged in slices in a similar way to MPEG-2 coded video [7]. Each slice includes a start code and a series of coded macroblocks and the slices are arranged in raster order to cover the coded picture (see for example Figure 5.73). This structure is adopted to simplify transcoding to/from an MPEG-2 coded representation.

VOL headers: Additional data fields are added to the VOL header, mimicking those in an MPEG-2 picture header in order to simplify MPEG-2 transcoding.

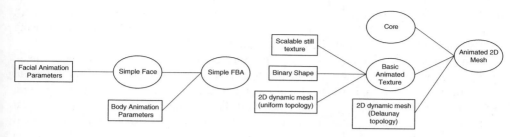

Figure 5.74   Tools and objects for animation

5.7.2 The Core Studio Profile

The Core Studio object is intended for distribution of studio-quality video (for example be-tween production studios) and adds support for Sprites and P-VOPs to the Simple Studio tools. Sprite coding is modified by adding extra sprite control parameters that closely mimic the properties of 'real' video cameras, such as lens distortion. Motion compensation and mo-tion vector coding in P-VOPs is modified for compatibility with the MPEG-2 syntax, for example, motion vectors are predictively coded using the MPEG-2 method rather than the usual MPEG-4 median prediction method.

5.8 CODING SYNTHETIC VISUAL SCENES

For the first time in an international standard, MPEG4 introduced the concept of 'hybrid' synthetic and natural video objects for visual communication. According to this concept, some applications may benefit from using a combination of tools from the video coding community (designed for coding of 'real world' or 'natural' video material) and tools from the 2D/3D animation community (designed for rendering 'synthetic' or computer-generated visual scenes).

MPEG4 Visual includes several tools and objects that can make use of a combination of animation and natural video processing (Figure 5.74). The *Basic Animated Texture* and *Animated 2D Mesh* object types support the coding of 2D meshes that represent shape and motion, together with still texture that may be mapped onto a mesh. A tool for representing and coding *3D Mesh* models is included in MPEG-4 Visual Version 2 but is not yet part of any profile. The *Face and Body Animation* tools enable a human face and/or body to be modelled and coded [8].

It has been shown that animation-based tools have potential applications to very low bit rate video coding [9]. However, in practice, the main application of these tools to date has been in coding synthetic (computer-generated) material. As the focus of this book is natural video coding, these tools will not be covered in detail.

5.8.1 Animated 2D and 3D Mesh Coding

A 2D mesh is made up of triangular patches and covers the 2D plane of an image or VO. De-formation or motion between VOPs can be modelled by warping the triangular patches. A 3D mesh models an object as a collection of polygons in 3D space (including depth information). The surface texture of a 2D or 3D mesh may be compressed as a *static texture* image (using the DWT) which is projected onto the mesh at the decoder.

These tools have a number of applications including coding and representing synthetic (computer generated) or 'natural' 2D and 3D objects in a scene and may be particularly useful for applications that combine natural and computer-generated imagery (Synthetic Natural Hybrid Coding, SNHC). It is possible to use mesh-based coding for compression of natural video objects. For example, a mesh and a static texture map may be transmitted for selected key frames. No texture is transmitted for intermediate frames, instead the mesh parameters are transmitted and the decoder reconstructs intermediate frames by animating (deforming) the mesh and the surface texture. This type of representation has much more flexibility than block-based motion models, since the mesh can adapt to deformations and nonlinear motion between frames. At the present time, however, automatic generation and tracking of mesh parameters is prohibitively complex for practical video compression applications.

The Basic Animated Texture object type includes the Binary Shape, Scalable Still Texture and 2D Dynamic Mesh tools and supports coding of 2D meshes (with uniform topology) and arbitrary-shaped still texture maps. The Animated 2D Mesh object type is a superset of Basic Animated Texture and adds Delaunay mesh topology (a more flexible method of defining mesh triangles) and the Core Profile video coding tools to enable a flexible combination of animated meshes and video objects.

5.8.2 Face and Body Animation

MPEG-4 Visual includes specific support for animated human face and body models within the Simple Face Animation and Simple FBA (Face and Body Animation) visual object types. The basic approach to face/body animation consists of (a) defining the geometric shape of a body or face model (typically carried out once at the start of a session) and (b) sending animation parameters to animate the body/face model.

A face model is described by Facial Definition Parameters (FDPs) and animated using Facial Animation Parameters (FAPs). The default set of FDPs may be used to render a 'generic' face at the decoder and a custom set of FDPs may be transmitted to create a model of a specific face. Once the model is available at the decoder, it can be animated by transmitting FAPs. In a similar way, a body object is rendered from a set of Body Definition Parameters (BDPs) and animated using Body Animation Parameters (BAPs). This enables the rendering of body models ranging from a generic, synthetic body to a body with a specific shape.

Applications for face and body animation include the generation of virtual scenes or worlds containing synthetic face and/or body objects, as well as model-based coding of natural face or body scenes, in which face and/or body movement is analysed, transmitted as a set of BAPs/FAPs and synthesized at the decoder. To date, these methods have not been widely used for video coding.

5.9 CONCLUSIONS

The MPEG-4 Visual standard supports the coding and representation of visual objects with efficient compression and unparalleled flexibility. The diverse set of coding tools described in the standard are capable of supporting a wide range of applications such as efficient coding of video frames, video coding for unreliable transmission networks, object-based coding and manipulation, coding of synthetic and 'hybrid' synthetic/natural scenes and highly interactive visual applications.

The MPEG-4 standard continues to develop with the addition of new tools (for example, profiles to support streaming video). However, amongst developers and manufacturers, the most popular elements of MPEG-4 Visual to date have been the Simple and Advanced Simple Profile tools and there is a clear industry requirement for efficient coding of rectangular video frames. This requirement, together with a protracted period of uncertainty about MPEG-4 Visual patent and licensing issues (see Chapter 8), means that the newly-developed H.264 standard is showing signs of overtaking MPEG-4 Visual in the market. The next chapter examines H.264 in detail.

5.10 REFERENCES

1. ISO/IEC 14496-2, Amendment 1, Information technology – coding of audio-visual objects – Part 2: Visual, 2001.
2. ISO/IEC 14496-1, Information technology – coding of audio-visual objects – Part 1: Systems, 2001
3. Y. Wang, S. Wenger, J. Wen and A. Katsaggelos, Review of error resilient coding techniques for real-time video communications, *IEEE Signal Process. Mag.*, July 2000.
4. N. Brady, MPEG-4 standardized methods for the compression of arbitrarily shaped video objects, *IEEE Trans. Circuits Syst. Video Technol.*, pp. 1170–1189, 1999.
5. W. Li, Overview of Fine Granular Scalability in MPEG-4 Video standard, *IEEE Trans. Circuits Syst. Video Technol.*, **11**(3), March 2001.
6. I. Daubechies, The wavelet transform, time-frequency localization and signal analysis, *IEEE Trans. Inf. Theory* 36, pp. 961–1005, 1990.
7. ISO/IEC 13818, Information technology: generic coding of moving pictures and associated audio information, 1995 (MPEG-2).
8. I. Pandzic and R. Forchheimer, *MPEG-4 Facial Animation*, John Wiley & Sons, August 2002.
9. P. Eisert, T. Wiegand, and B. Girod, Model-aided coding: a new approach to incorporate facial animation into motion-compensated video coding, *IEEE Trans. Circuits Syst. Video Technol.*, **10**(3), pp. 344–358, April 2000.

6

H.264/MPEG4 Part 10

6.1 INTRODUCTION

The Moving Picture Experts Group and the Video Coding Experts Group (MPEG and VCEG) have developed a new standard that promises to outperform the earlier MPEG-4 and H.263 standards, providing better compression of video images. The new standard is entitled 'Advanced Video Coding' (AVC) and is published jointly as Part 10 of MPEG-4 and ITU-T Recommendation H.264 [1, 3].

6.1.1 Terminology

Some of the important terminology adopted in the H.264 standard is as follows (the details of these concepts are explained in later sections):

A field (of interlaced video) or a frame (of progressive or interlaced video) is encoded to produce a *coded picture*. A coded frame has a *frame number* (signalled in the bitstream), which is not necessarily related to decoding order, and each coded field of a progressive or interlaced frame has an associated *picture order count*, which defines the decoding order of fields. Previously coded pictures (*reference pictures*) may be used for inter prediction of further coded pictures. Reference pictures are organised into one or two lists (sets of numbers corresponding to reference pictures), described as *list 0* and *list 1*.

A coded picture consists of a number of *macroblocks*, each containing 16×16 luma samples and associated chroma samples (8×8 Cb and 8×8 Cr samples in the current standard). Within each picture, macroblocks are arranged in *slices*, where a slice is a set of macroblocks in raster scan order (but not necessarily contiguous – see section 6.4.3). An *I slice* may contain only I macroblock types (see below), a *P slice* may contain P and I macroblock types and a *B slice* may contain B and I macroblock types. (There are two further slice types, SI and SP, discussed in section 6.6.1).

I macroblocks are predicted using intra prediction from decoded samples in the current slice. A prediction is formed either (a) for the complete macroblock or (b) for each 4×4 block

H.264 and MPEG-4 Video Compression: Video Coding for Next-generation Multimedia.
Iain E. G. Richardson.   © 2003 John Wiley & Sons, Ltd. ISBN: 0-470-84837-5

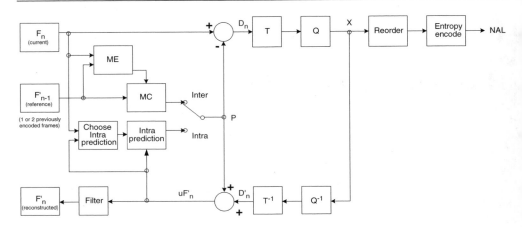

Figure 6.1   H.264 Encoder

of luma samples (and associated chroma samples) in the macroblock. (An alternative to intra prediction, I_PCM mode, is described in section 6.4.6).

P macroblocks are predicted using inter prediction from reference picture(s). An inter coded macroblock may be divided into *macroblock partitions*, i.e. blocks of size 16×16, 16×8, 8×16 or 8×8 luma samples (and associated chroma samples). If the 8×8 partition size is chosen, each 8×8 *sub-macroblock* may be further divided into *sub-macroblock partitions* of size 8×8, 8×4, 4×8 or 4×4 luma samples (and associated chroma samples). Each macroblock partition may be predicted from one picture in list 0. If present, every sub-macroblock partition in a sub-macroblock is predicted from the same picture in list 0.

B macroblocks are predicted using inter prediction from reference picture(s). Each macroblock partition may be predicted from one or two reference pictures, one picture in list 0 and/or one picture in list 1. If present, every sub-macroblock partition in a sub-macroblock is predicted from (the same) one or two reference pictures, one picture in list 0 and/or one picture in list 1.

6.2 THE H.264 CODEC

In common with earlier coding standards, H.264 does not explicitly define a CODEC (enCOder / DECoder pair) but rather defines the syntax of an encoded video bitstream together with the method of decoding this bitstream. In practice, a compliant encoder and decoder are likely to include the functional elements shown in Figure 6.1 and Figure 6.2. With the exception of the deblocking filter, most of the basic functional elements (prediction, transform, quantisation, entropy encoding) are present in previous standards (MPEG-1, MPEG-2, MPEG-4, H.261, H.263) but the important changes in H.264 occur in the details of each functional block.

The Encoder (Figure 6.1) includes two dataflow paths, a 'forward' path (left to right) and a 'reconstruction' path (right to left). The dataflow path in the Decoder (Figure 6.2) is shown from right to left to illustrate the similarities between Encoder and Decoder. Before examining the detail of H.264, we will describe the main steps in encoding and decoding a frame (or field)

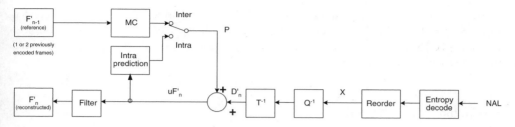

Figure 6.2   H.264 Decoder

of video. The following description is simplified in order to provide an overview of encoding and decoding. The term "block" is used to denote a macroblock partition or sub-macroblock partition (inter coding) or a 16×16 or 4×4 block of luma samples and associated chroma samples (intra coding).

Encoder (forward Path)

An input frame or field F_n is processed in units of a macroblock. Each macroblock is encoded in intra or inter mode and, for each block in the macroblock, a prediction PRED (marked 'P' in Figure 6.1) is formed based on reconstructed picture samples. In Intra mode, PRED is formed from samples in the current slice that have previously encoded, decoded and reconstructed (uF$'_n$ in the figures; note that *unfiltered* samples are used to form PRED). In Inter mode, PRED is formed by motion-compensated prediction from one or two reference picture(s) selected from the set of list 0 and/or list 1 reference pictures. In the figures, the reference picture is shown as the previous encoded picture F'_{n-1} but the prediction reference for each macroblock partition (in inter mode) may be chosen from a selection of past or future pictures (in display order) that have already been encoded, reconstructed and filtered.

The prediction PRED is subtracted from the current block to produce a residual (difference) block D_n that is transformed (using a block transform) and quantised to give X, a set of quantised transform coefficients which are reordered and entropy encoded. The entropy-encoded coefficients, together with side information required to decode each block within the macroblock (prediction modes, quantiser parameter, motion vector information, etc.) form the compressed bitstream which is passed to a Network Abstraction Layer (NAL) for transmission or storage.

Encoder (Reconstruction Path)

As well as encoding and transmitting each block in a macroblock, the encoder decodes (reconstructs) it to provide a reference for further predictions. The coefficients X are scaled (Q^{-1}) and inverse transformed (T^{-1}) to produce a difference block D'_n. The prediction block PRED is added to D'_n to create a reconstructed block uF$'_n$ (a decoded version of the original block; u indicates that it is unfiltered). A filter is applied to reduce the effects of blocking distortion and the reconstructed reference picture is created from a series of blocks F'_n.

Decoder

The decoder receives a compressed bitstream from the NAL and entropy decodes the data elements to produce a set of quantised coefficients X. These are scaled and inverse transformed to give D'_n (identical to the D'_n shown in the Encoder). Using the header information decoded from the bitstream, the decoder creates a prediction block PRED, identical to the original prediction PRED formed in the encoder. PRED is added to D'_n to produce uF'_n which is filtered to create each decoded block F'_n.

6.3 H.264 STRUCTURE

6.3.1 Profiles and Levels

H.264 defines a set of three *Profiles*, each supporting a particular set of coding functions and each specifying what is required of an encoder or decoder that complies with the Profile. The *Baseline Profile* supports intra and inter-coding (using I-slices and P-slices) and entropy coding with context-adaptive variable-length codes (CAVLC). The *Main Profile* includes support for interlaced video, inter-coding using B-slices, inter coding using weighted prediction and entropy coding using context-based arithmetic coding (CABAC). The *Extended Profile* does not support interlaced video or CABAC but adds modes to enable efficient switching between coded bitstreams (SP- and SI-slices) and improved error resilience (Data Partitioning). Potential applications of the Baseline Profile include videotelephony, videoconferencing and wireless communications; potential applications of the Main Profile include television broadcasting and video storage; and the Extended Profile may be particularly useful for streaming media applications. However, each Profile has sufficient flexibility to support a wide range of applications and so these examples of applications should not be considered definitive.

Figure 6.3 shows the relationship between the three Profiles and the coding tools supported by the standard. It is clear from this figure that the Baseline Profile is a subset of the Extended Profile, but not of the Main Profile. The details of each coding tool are described in Sections 6.4, 6.5 and 6.6 (starting with the Baseline Profile tools).

Performance limits for CODECs are defined by a set of Levels, each placing limits on parameters such as sample processing rate, picture size, coded bitrate and memory requirements.

6.3.2 Video Format

H.264 supports coding and decoding of 4:2:0 progressive or interlaced video[1] and the default sampling format for 4:2:0 progressive frames is shown in Figure 2.11 (other sampling formats may be signalled as Video Usability Information parameters). In the default sampling format, chroma (Cb and Cr) samples are aligned horizontally with every 2nd luma sample and are located vertically between two luma samples. An interlaced frame consists of two fields (a top field and a bottom field) separated in time and with the default sampling format shown in Figure 2.12.

---

[1] An extension to H.264 to support alternative colour sampling structures and higher sample accuracy is currently under development.

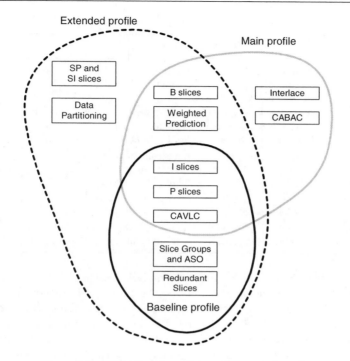

Figure 6.3 H.264 Baseline, Main and Extended profiles

NAL header	RBSP	NAL header	RBSP	NAL header	RBSP

Figure 6.4 Sequence of NAL units

6.3.3 Coded Data Format

H.264 makes a distinction between a Video Coding Layer (VCL) and a Network Abstraction Layer (NAL). The output of the encoding process is VCL data (a sequence of bits representing the coded video data) which are mapped to NAL units prior to transmission or storage. Each NAL unit contains a Raw Byte Sequence Payload (RBSP), a set of data corresponding to coded video data or header information. A coded video sequence is represented by a sequence of NAL units (Figure 6.4) that can be transmitted over a packet-based network or a bitstream transmission link or stored in a file. The purpose of separately specifying the VCL and NAL is to distinguish between coding-specific features (at the VCL) and transport-specific features (at the NAL). Section 6.7 describes the NAL and transport mechanisms in more detail.

6.3.4 Reference Pictures

An H.264 encoder may use one or two of a number of previously encoded pictures as a reference for motion-compensated prediction of each inter coded macroblock or macroblock

Table 6.1 H.264 slice modes

Slice type	Description	Profile(s)
I (Intra)	Contains only I macroblocks (each block or macroblock is predicted from previously coded data within the same slice).	All
P (Predicted)	Contains P macroblocks (each macroblock or macroblock partition is predicted from one list 0 reference picture) and/or I macroblocks.	All
B (Bi-predictive)	Contains B macroblocks (each macroblock or macroblock partition is predicted from a list 0 and/or a list 1 reference picture) and/or I macroblocks.	Extended and Main
SP (Switching P)	Facilitates switching between coded streams; contains P and/or I macroblocks.	Extended
SI (Switching I)	Facilitates switching between coded streams; contains SI macroblocks (a special type of intra coded macroblock).	Extended

partition. This enables the encoder to search for the best 'match' for the current macroblock partition from a wider set of pictures than just (say) the previously encoded picture.

The encoder and decoder each maintain one or two lists of reference pictures, containing pictures that have previously been encoded and decoded (occurring before and/or after the current picture in display order). Inter coded macroblocks and macroblock partitions in P slices (see below) are predicted from pictures in a single list, **list 0**. Inter coded macroblocks and macroblock partitions in a B slice (see below) may be predicted from two lists, **list 0** and **list 1**.

6.3.5 Slices

A video picture is coded as one or more slices, each containing an integral number of macroblocks from 1 (1 MB per slice) to the total number of macroblocks in a picture (1 slice per picture) The number of macroblocks per slice need not be constant within a picture. There is minimal inter-dependency between coded slices which can help to limit the propagation of errors. There are five types of coded slice (Table 6.1) and a coded picture may be composed of different types of slices. For example, a Baseline Profile coded picture may contain a mixture of I and P slices and a Main or Extended Profile picture may contain a mixture of I, P and B slices.

Figure 6.5 shows a simplified illustration of the syntax of a coded slice. The slice header defines (among other things) the slice type and the coded picture that the slice 'belongs' to and may contain instructions related to reference picture management (see Section 6.4.2). The slice data consists of a series of coded macroblocks and/or an indication of skipped (not coded) macroblocks. Each MB contains a series of header elements (see Table 6.2) and coded residual data.

6.3.6 Macroblocks

A macroblock contains coded data corresponding to a 16×16 sample region of the video frame (16×16 luma samples, 8×8 Cb and 8×8 Cr samples) and contains the syntax elements described in Table 6.2. Macroblocks are numbered (addressed) in raster scan order within a frame.

Table 6.2   Macroblock syntax elements

mb_type	Determines whether the macroblock is coded in intra or inter (P or B) mode; determines macroblock partition size (see Section 6.4.2).
mb_pred	Determines intra prediction modes (intra macroblocks); determines list 0 and/or list 1 references and differentially coded motion. vectors for each macroblock partition (inter macroblocks, except for inter MBs with 8 × 8 macroblock partition size).
sub_mb_pred	(Inter MBs with 8 × 8 macroblock partition size only) Determines sub-macroblock partition size for each sub-macroblock; list 0 and/or list 1 references for each macroblock partition; differentially coded motion vectors for each macroblock sub-partition.
coded_block_pattern	Identifies which 8 × 8 blocks (luma and chroma) contain coded transform coefficients.
mb_qp_delta	Changes the quantiser parameter (see Section 6.4.8).
residual	Coded transform coefficients corresponding to the residual image samples after prediction (see Section 6.4.8).

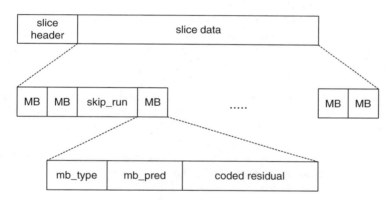

Figure 6.5   Slice syntax

6.4 THE BASELINE PROFILE

6.4.1 Overview

The Baseline Profile supports coded sequences containing I- and P-slices. I-slices contain intra-coded macroblocks in which each 16 × 16 or 4 × 4 luma region and each 8 × 8 chroma region is predicted from previously-coded samples in the same slice. P-slices may contain intra-coded, inter-coded or skipped MBs. Inter-coded MBs in a P slice are predicted from a number of previously coded pictures, using motion compensation with quarter-sample (luma) motion vector accuracy.

After prediction, the residual data for each MB is transformed using a 4 × 4 integer transform (based on the DCT) and quantised. Quantised transform coefficients are reordered and the syntax elements are entropy coded. In the Baseline Profile, transform coefficients are entropy coded using a context-adaptive variable length coding scheme (CAVLC) and all other syntax elements are coded using fixed-length or Exponential-Golomb Variable Length Codes. Quantised coefficients are scaled, inverse transformed, reconstructed (added to the prediction formed during encoding) and filtered with a de-blocking filter before (optionally) being stored for possible use in reference pictures for further intra- and inter-coded macroblocks.

6.4.2 Reference Picture Management

Pictures that have previously been encoded are stored in a reference buffer (the decoded picture buffer, DPB) in both the encoder and the decoder. The encoder and decoder maintain a list of previously coded pictures, reference picture list 0, for use in motion-compensated prediction of inter macroblocks in P slices. For P slice prediction, list 0 can contain pictures before and after the current picture in display order and may contain both *short term* and *long term* reference pictures. By default, an encoded picture is reconstructed by the encoder and marked as a short term picture, a recently-coded picture that is available for prediction. Short term pictures are identified by their frame number. Long term pictures are (typically) older pictures that may also be used for prediction and are identified by a variable LongTermPicNum. Long term pictures remain in the DPB until explicitly removed or replaced.

When a picture is encoded and reconstructed (in the encoder) or decoded (in the decoder), it is placed in the decoded picture buffer and is either (a) marked as 'unused for reference' (and hence not used for any further prediction), (b) marked as a short term picture, (c) marked as a long term picture or (d) simply output to the display. By default, short term pictures in list 0 are ordered from the highest to the lowest PicNum (a variable derived from the frame number) and long term pictures are ordered from the lowest to the highest LongTermPicNum. The encoder may signal a change to the default reference picture list order. As each new picture is added to the short term list at position 0, the indices of the remaining short-term pictures are incremented. If the number of short term and long term pictures is equal to the maximum number of reference frames, the oldest short-term picture (with the highest index) is removed from the buffer (known as *sliding window* memory control). The effect that this process is that the encoder and decoder each maintain a 'window' of N short-term reference pictures, including the current picture and $(N - 1)$ previously encoded pictures.

Adaptive memory control commands, sent by the encoder, manage the short and long term picture indexes. Using these commands, a short term picture may be assigned a long term frame index, or any short term or long term picture may be marked as 'unused for reference'.

The encoder chooses a reference picture from list 0 for encoding each macroblock partition in an inter-coded macroblock. The choice of reference picture is signalled by an index number, where index 0 corresponds to the first frame in the short term section and the indices of the long term frames start after the last short term frame (as shown in the following example).

Example: Reference buffer management (P-slice)

Current frame number = 250

Number of reference frames = 5

Operation	Reference picture list				
	0	1	2	3	4
Initial state	–	–	–	–	–
Encode frame 250	250	–	–	–	–
Encode 251	251	250	–	–	–
Encode 252	252	251	250	–	–
Encode 253	253	252	251	250	–
Assign 251 to LongTermPicNum 0	253	252	250	0	–
Encode 254	254	253	252	250	0
Assign 253 to LongTermPicNum 4	254	252	250	0	4
Encode 255	255	254	252	0	4
Assign 255 to LongTermPicNum 3	254	252	0	3	4
Encode 256	256	254	0	3	4

(Note that in the above example, 0, 3 and 4 correspond to the decoded frames 251, 255 and 253 respectively).

Instantaneous Decoder Refresh Picture

An encoder sends an IDR (Instantaneous Decoder Refresh) coded picture (made up of I- or SI-slices) to clear the contents of the reference picture buffer. On receiving an IDR coded picture, the decoder marks all pictures in the reference buffer as 'unused for reference'. All subsequent transmitted slices can be decoded without reference to any frame decoded prior to the IDR picture. The first picture in a coded video sequence is always an IDR picture.

6.4.3 Slices

A bitstream conforming to the the Baseline Profile contains coded I and/or P slices. An I slice contains only intra-coded macroblocks (predicted from previously coded samples in the same slice, see Section 6.4.6) and a P slice can contain inter coded macroblocks (predicted from samples in previously coded pictures, see Section 6.4.5), intra coded macroblocks or Skipped macroblocks. When a Skipped macroblock is signalled in the bitstream, no further data is sent for that macroblock. The decoder calculates a vector for the skipped macroblock (see Section 6.4.5.3) and reconstructs the macroblock using motion-compensated prediction from the first reference picture in list 0.

An H.264 encoder may optionally insert a picture delimiter RBSP unit at the boundary between coded pictures. This indicates the start of a new coded picture and indicates which slice types are allowed in the following coded picture. If the picture delimiter is not used, the decoder is expected to detect the occurrence of a new picture based on the header of the first slice in the new picture.

Redundant coded picture

A picture marked as 'redundant' contains a redundant representation of part or all of a coded picture. In normal operation, the decoder reconstructs the frame from 'primary'

Table 6.3 Macroblock to slice group map types

Type	Name	Description
0	Interleaved	run_length MBs are assigned to each slice group in turn (Figure 6.6).
1	Dispersed	MBs in each slice group are dispersed throughout the picture (Figure 6.7).
2	Foreground and background	All but the last slice group are defined as rectangular regions within the picture. The last slice group contains all MBs not contained in any other slice group (the 'background'). In the example in Figure 6.8, group 1 overlaps group 0 and so MBs not already allocated to group 0 are allocated to group 1.
3	Box-out	A 'box' is created starting from the centre of the frame (with the size controlled by encoder parameters) and containing group 0; all other MBs are in group 1 (Figure 6.9).
4	Raster scan	Group 0 contains MBs in raster scan order from the top-left and all other MBs are in group 1 (Figure 6.9).
5	Wipe	Group 0 contains MBs in vertical scan order from the top-left and all other MBs are in group 1 (Figure 6.9).
6	Explicit	A parameter, slice_group_id, is sent for each MB to indicate its slice group (i.e. the macroblock map is entirely user-defined).

(nonredundant)' pictures and discards any redundant pictures. However, if a primary coded picture is damaged (e.g. due to a transmission error), the decoder may replace the damaged area with decoded data from a redundant picture if available.

Arbitrary Slice Order (ASO)

The Baseline Profile supports Arbitrary Slice Order which means that slices in a coded frame may follow any decoding order. ASO is defined to be in use if the first macroblock in any slice in a decoded frame has a smaller macroblock address than the first macroblock in a *previously* decoded slice in the same picture.

Slice Groups

A *slice group* is a subset of the macroblocks in a coded picture and may contain one or more slices. Within each slice in a slice group, MBs are coded in raster order. If only one slice group is used per picture, then all macroblocks in the picture are coded in raster order (unless ASO is in use, see above). Multiple slice groups (described in previous versions of the draft standard as Flexible Macroblock Ordering or FMO) make it possible to map the sequence of coded MBs to the decoded picture in a number of flexible ways. The allocation of macroblocks is determined by a *macroblock to slice group map* that indicates which slice group each MB belongs to. Table 6.3 lists the different types of macroblock to slice group maps.

Example: 3 slice groups are used and the map type is 'interleaved' (Figure 6.6). The coded picture consists of first, all of the macroblocks in slice group 0 (filling every 3rd row of macroblocks); second, all of the macroblocks in slice group 1; and third, all of the macroblocks in slice group 0. Applications of multiple slice groups include error resilience, for example if one of the slice groups in the dispersed map shown in Figure 6.7 is 'lost' due to errors, the missing data may be concealed by interpolation from the remaining slice groups.

0
1
2
0
1
2
0
1
2

Figure 6.6   Slice groups: Interleaved map (QCIF, three slice groups)

0	1	2	3	0	1	2	3	0	1	2
2	3	0	1	2	3	0	1	2	3	0
0	1	2	3	0	1	2	3	0	1	2
2	3	0	1	2	3	0	1	2	3	0
0	1	2	3	0	1	2	3	0	1	2
2	3	0	1	2	3	0	1	2	3	0
0	1	2	3	0	1	2	3	0	1	2
2	3	0	1	2	3	0	1	2	3	0
0	1	2	3	0	1	2	3	0	1	2

Figure 6.7   Slice groups: Dispersed map (QCIF, four slice groups)

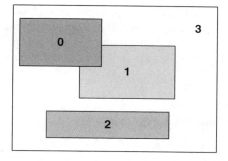

Figure 6.8   Slice groups: Foreground and Background map (four slice groups)

6.4.4 Macroblock Prediction

Every coded macroblock in an H.264 slice is predicted from previously-encoded data. Samples within an intra macroblock are predicted from samples in the current slice that have already been encoded, decoded and reconstructed; samples in an inter macroblock are predicted from previously-encoded.

A prediction for the current macroblock or block (a model that resembles the current macroblock or block as closely as possible) is created from image samples that have already

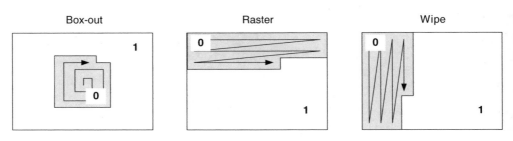

Figure 6.9   Slice groups: Box-out, Raster and Wipe maps

been encoded (either in the same slice or in a previously encoded slice). This prediction is subtracted from the current macroblock or block and the result of the subtraction (residual) is compressed and transmitted to the decoder, together with information required for the decoder to repeat the prediction process (motion vector(s), prediction mode, etc.). The decoder creates an identical prediction and adds this to the decoded residual or block. The encoder bases its prediction on encoded and decoded image samples (rather than on original video frame samples) in order to ensure that the encoder and decoder predictions are identical.

6.4.5 Inter Prediction

Inter prediction creates a prediction model from one or more previously encoded video frames or fields using block-based motion compensation. Important differences from earlier standards include the support for a range of block sizes (from 16 × 16 down to 4 × 4) and fine sub-sample motion vectors (quarter-sample resolution in the luma component). In this section we describe the inter prediction tools available in the Baseline profile. Extensions to these tools in the Main and Extended profiles include B-slices (Section 6.5.1) and Weighted Prediction (Section 6.5.2).

6.4.5.1 Tree structured motion compensation

The luminance component of each macroblock (16 × 16 samples) may be split up in four ways (Figure 6.10) and motion compensated either as one 16 × 16 *macroblock partition*, two 16 × 8 partitions, two 8 × 16 partitions or four 8 × 8 partitions. If the 8 × 8 mode is chosen, each of the four 8 × 8 sub-macroblocks within the macroblock may be split in a further 4 ways (Figure 6.11), either as one 8 × 8 sub-macroblock partition, two 8 × 4 sub-macroblock partitions, two 4 × 8 sub-macroblock partitions or four 4 × 4 sub-macroblock partitions. These partitions and sub-macroblock give rise to a large number of possible combinations within each macroblock. This method of partitioning macroblocks into motion compensated sub-blocks of varying size is known as *tree structured motion compensation*.

A separate motion vector is required for each partition or sub-macroblock. Each motion vector must be coded and transmitted and the choice of partition(s) must be encoded in the compressed bitstream. Choosing a large partition size (16 × 16, 16 × 8, 8 × 16) means that

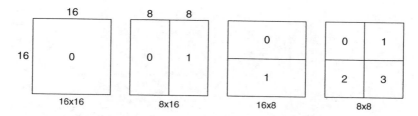

Figure 6.10 Macroblock partitions: 16×16, 8×16, 16×8, 8×8

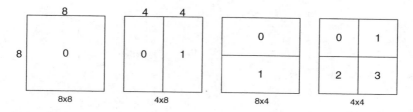

Figure 6.11 Sub-macroblock partitions: 8×8, 4×8, 8×4, 4×4

a small number of bits are required to signal the choice of motion vector(s) and the type of partition but the motion compensated residual may contain a significant amount of energy in frame areas with high detail. Choosing a small partition size (8×4, 4×4, etc.) may give a lower-energy residual after motion compensation but requires a larger number of bits to signal the motion vectors and choice of partition(s). The choice of partition size therefore has a significant impact on compression performance. In general, a large partition size is appropriate for homogeneous areas of the frame and a small partition size may be beneficial for detailed areas.

Each chroma component in a macroblock (Cb and Cr) has half the horizontal and vertical resolution of the luminance (luma) component. Each chroma block is partitioned in the same way as the luma component, except that the partition sizes have exactly half the horizontal and vertical resolution (an 8×16 partition in luma corresponds to a 4×8 partition in chroma; an 8×4 partition in luma corresponds to 4×2 in chroma and so on). The horizontal and vertical components of each motion vector (one per partition) are halved when applied to the chroma blocks.

Example

Figure 6.12 shows a residual frame (without motion compensation). The H.264 reference encoder selects the 'best' partition size for each part of the frame, in this case the partition size that minimises the amount of information to be sent, and the chosen partitions are shown superimposed on the residual frame. In areas where there is little change between the frames (residual appears grey), a 16×16 partition is chosen and in areas of detailed motion (residual appears black or white), smaller partitions are more efficient.

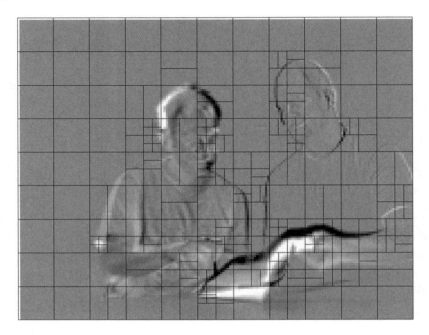

Figure 6.12   Residual (without MC) showing choice of block sizes

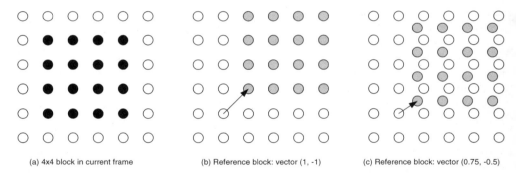

(a) 4x4 block in current frame   (b) Reference block: vector (1, -1)   (c) Reference block: vector (0.75, -0.5)

Figure 6.13   Example of integer and sub-sample prediction

6.4.5.2 Motion Vectors

Each partition or sub-macroblock partition in an inter-coded macroblock is predicted from an area of the same size in a reference picture. The offset between the two areas (the motion vector) has quarter-sample resolution for the luma component and one-eighth-sample resolution for the chroma components. The luma and chroma samples at sub-sample positions do not exist in the reference picture and so it is necessary to create them using interpolation from nearby coded samples. In Figure 6.13, a 4×4 block in the current frame (a) is predicted from a region of the reference picture in the neighbourhood of the current block position. If the horizontal and vertical components of the motion vector are integers (b), the relevant samples in the

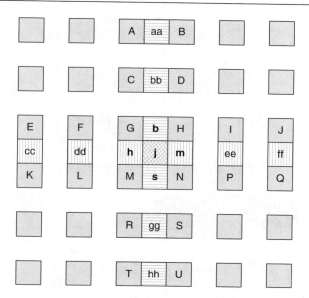

Figure 6.14 Interpolation of luma half-pel positions

reference block actually exist (grey dots). If one or both vector components are fractional values (c), the prediction samples (grey dots) are generated by interpolation between adjacent samples in the reference frame (white dots).

Generating Interpolated Samples

The samples half-way between integer-position samples ('half-pel samples') in the luma component of the reference picture are generated first (Figure 6.14, grey markers). Each half-pel sample that is adjacent to two integer samples (e.g. b, h, m, s in Figure 6.14) is interpolated from integer-position samples using a six tap Finite Impulse Response (FIR) filter with weights $(1/32, -5/32, 5/8, 5/8, -5/32, 1/32)$. For example, half-pel sample **b** is calculated from the six horizontal integer samples E, F, G, H, I and J:

$$\mathbf{b} = \text{round}((E - 5F + 20G + 20H - 5I + J) /32)$$

Similarly, **h** is interpolated by filtering A, C, G, M, R and T. Once all of the samples horizontally and vertically adjacent to integer samples have been calculated, the remaining half-pel positions are calculated by interpolating between six horizontal or vertical half-pel samples from the first set of operations. For example, **j** is generated by filtering cc, dd, h, m, ee and ff (note that the result is the same whether **j** is interpolated horizontally or vertically; note also that un-rounded versions of h and m are used to generate j). The six-tap interpolation filter is relatively complex but produces an accurate fit to the integer-sample data and hence good motion compensation performance.

Once all the half-pel samples are available, the samples at quarter-step ('quarter-pel') positions are produced by linear interpolation (Figure 6.15). Quarter-pel positions with two horizontally or vertically adjacent half- or integer-position samples (e.g. **a, c, i, k** and **d, f, n,**

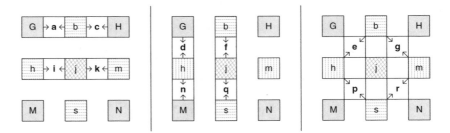

Figure 6.15   Interpolation of luma quarter-pel positions

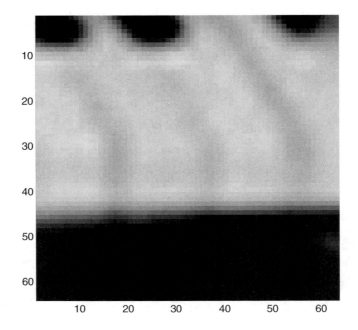

Figure 6.16   Luma region interpolated to quarter-pel positions

q in Figure 6.15) are linearly interpolated between these adjacent samples, for example:

$$a = \text{round}((G + b) / 2)$$

The remaining quarter-pel positions (**e, g, p** and **r** in the figure) are linearly interpolated between a pair of diagonally opposite *half*-pel samples. For example, **e** is interpolated between b and h. Figure 6.16 shows the result of interpolating the reference region shown in Figure 3.16 with quarter-pel resolution.

Quarter-pel resolution motion vectors in the luma component require eighth-sample resolution vectors in the chroma components (assuming 4:2:0 sampling). Interpolated samples are generated at eighth-sample intervals between integer samples in each chroma component using linear interpolation (Figure 6.17). Each sub-sample position **a** is a linear combination

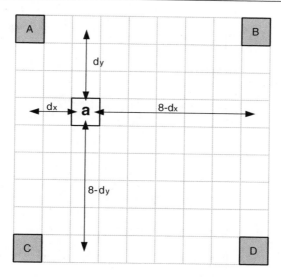

Figure 6.17 Interpolation of chroma eighth-sample positions

of the neighbouring integer sample positions A, B, C and D:

$$\mathbf{a} = \text{round}([(8 - d_x) \cdot (8 - d_y)A + d_x \cdot (8 - d_y)B + (8 - d_x) \cdot d_yC + d_x \cdot d_yD]/64)$$

In Figure 6.17, d_x is 2 and d_y is 3, so that:

$$\mathbf{a} = \text{round}[(30A + 10B + 18C + 6D)/64]$$

6.4.5.3 Motion Vector Prediction

Encoding a motion vector for each partition can cost a significant number of bits, especially if small partition sizes are chosen. Motion vectors for neighbouring partitions are often highly correlated and so each motion vector is predicted from vectors of nearby, previously coded partitions. A predicted vector, MVp, is formed based on previously calculated motion vectors and MVD, the difference between the current vector and the predicted vector, is encoded and transmitted. The method of forming the prediction MVp depends on the motion compensation partition size and on the availability of nearby vectors.

Let E be the current macroblock, macroblock partition or sub-macroblock partition, let A be the partition or sub-partition immediately to the left of E, let B be the partition or sub-partition immediately above E and let C be the partition or sub-macroblock partition above and to the right of E. If there is more than one partition immediately to the left of E, the topmost of these partitions is chosen as A. If there is more than one partition immediately above E, the leftmost of these is chosen as B. Figure 6.18 illustrates the choice of neighbouring partitions when all the partitions have the same size (16×16 in this case) and Figure 6.19 shows an

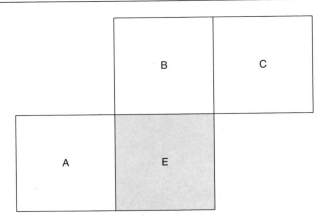

Figure 6.18   Current and neighbouring partitions (same partition sizes)

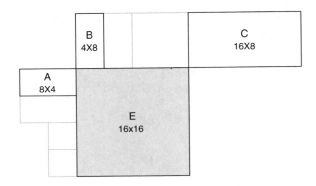

Figure 6.19   Current and neighbouring partitions (different partition sizes)

example of the choice of prediction partitions when the neighbouring partitions have different sizes from the current partition E.

1. For transmitted partitions excluding 16 × 8 and 8 × 16 partition sizes, MVp is the median of the motion vectors for partitions A, B and C.
2. For 16 × 8 partitions, MVp for the upper 16 × 8 partition is predicted from B and MVp for the lower 16 × 8 partition is predicted from A.
3. For 8 × 16 partitions, MVp for the left 8 × 16 partition is predicted from A and MVp for the right 8 × 16 partition is predicted from C.
4. For skipped macroblocks, a 16 × 16 vector MVp is generated as in case (1) above (i.e. as if the block were encoded in 16 × 16 Inter mode).

If one or more of the previously transmitted blocks shown in Figure 6.19 is not available (e.g. if it is outside the current slice), the choice of MVp is modified accordingly. At the decoder, the predicted vector MVp is formed in the same way and added to the decoded vector difference MVD. In the case of a skipped macroblock, there is no decoded vector difference and a motion-compensated macroblock is produced using MVp as the motion vector.

Figure 6.20   QCIF frame

6.4.6 Intra Prediction

In intra mode a prediction block P is formed based on previously encoded and reconstructed blocks and is subtracted from the current block prior to encoding. For the luma samples, P is formed for each 4 × 4 block or for a 16 × 16 macroblock. There are a total of nine optional prediction modes for each 4 × 4 luma block, four modes for a 16 × 16 luma block and four modes for the chroma components. The encoder typically selects the prediction mode for each block that minimises the difference between P and the block to be encoded.

> ### Example
>
> A QCIF video frame (Figure 6.20) is encoded in intra mode and each block or macroblock is predicted from neighbouring, previously-encoded samples. Figure 6.21 shows the predicted luma frame P formed by choosing the 'best' 4 × 4 or 16 × 16 prediction mode for each region (the mode that minimises the amount of information to be coded).

A further intra coding mode, I_PCM, enables an encoder to transmit the values of the image samples directly (without prediction or transformation). In some special cases (e.g. anomalous image content and/or very low quantizer parameters), this mode may be more efficient than the 'usual' process of intra prediction, transformation, quantization and entropy coding. Including the I_PCM option makes it possible to place an absolute limit on the number of bits that may be contained in a coded macroblock without constraining decoded image quality.

6.4.6.1 4 × 4 Luma Prediction Modes

Figure 6.22 shows a 4 × 4 luma block (part of the highlighted macroblock in Figure 6.20) that is required to be predicted. The samples above and to the left (labelled A–M in Figure 6.23)

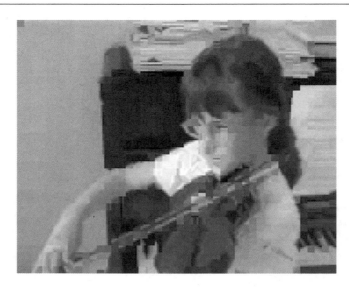

Figure 6.21   Predicted luma frame formed using H.264 intra prediction

4x4 luma block to be predicted

Figure 6.22   4 × 4 luma block to be predicted

have previously been encoded and reconstructed and are therefore available in the encoder
and decoder to form a prediction reference. The samples a, b, c, . . . , p of the prediction block
P (Figure 6.23) are calculated based on the samples A–M as follows. Mode 2 (DC prediction)
is modified depending on which samples A–M have previously been coded; each of the other
modes may only be used if all of the required prediction samples are available[2].

---

[2] Note that if samples E, F, G and H have not yet been decoded, the value of sample D is copied to these positions
and they are marked as 'available'.

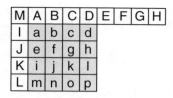

Figure 6.23 Labelling of prediction samples (4 × 4)

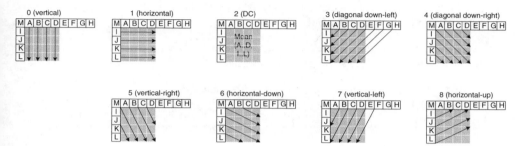

Figure 6.24 4 × 4 luma prediction modes

Mode 0 (Vertical)	The upper samples A, B, C, D are extrapolated vertically.
Mode 1 (Horizontal)	The left samples I, J, K, L are extrapolated horizontally.
Mode 2 (DC)	All samples in P are predicted by the mean of samples A . . . D and I . . . L.
Mode 3 (Diagonal Down-Left)	The samples are interpolated at a 45° angle between lower-left and upper-right.
Mode 4 (Diagonal Down-Right)	The samples are extrapolated at a 45° angle down and to the right.
Mode 5 (Vertical-Right)	Extrapolation at an angle of approximately 26.6° to the left of vertical (width/height = 1/2).
Mode 6 (Horizontal-Down)	Extrapolation at an angle of approximately 26.6° below horizontal.
Mode 7 (Vertical-Left)	Extrapolation (or interpolation) at an angle of approximately 26.6° to the right of vertical.
Mode 8 (Horizontal-Up)	Interpolation at an angle of approximately 26.6° above horizontal.

The arrows in Figure 6.24 indicate the direction of prediction in each mode. For modes 3–8, the predicted samples are formed from a *weighted average* of the prediction samples A–M. For example, if mode 4 is selected, the top-right sample of P (labelled 'd' in Figure 6.23) is predicted by: round(B/4+ C/2+ D/4).

Example

The nine prediction modes (0–8) are calculated for the 4 × 4 block shown in Figure 6.22 and the resulting prediction blocks P are shown in Figure 6.25. The Sum of Absolute Errors (SAE) for each prediction indicates the magnitude of the prediction error. In this case, the best match to the actual current block is given by mode 8 (horizontal-up) because this mode gives the smallest SAE and a visual comparison shows that the P block appears quite similar to the original 4 × 4 block.

6.4.6.2 16 × 16 Luma Prediction Modes

As an alternative to the 4 × 4 luma modes described in the previous section, the entire 16 × 16 luma component of a macroblock may be predicted in one operation. Four modes are available, shown in diagram form in Figure 6.26:

Mode 0 (vertical)	Extrapolation from upper samples (H)
Mode 1 (horizontal)	Extrapolation from left samples (V)
Mode 2 (DC)	Mean of upper and left-hand samples (H + V).
Mode 4 (Plane)	A linear 'plane' function is fitted to the upper and left-hand samples H and V. This works well in areas of smoothly-varying luminance.

Example:

Figure 6.27 shows a luminance macroblock with previously-encoded samples at the upper and left-hand edges. The results of the four prediction modes, shown in Figure 6.28, indicate that the best match is given by mode 3 which in this case produces a plane with a luminance gradient from light (upper-left) to dark (lower-right). Intra 16 × 16 mode works best in homogeneous areas of an image.

6.4.6.3 8 × 8 Chroma Prediction Modes

Each 8 × 8 chroma component of an intra coded a macroblock is predicted from previously encoded chroma samples above and/or to the left and both chroma components always use the same prediction mode. The four prediction modes are very similar to the 16 × 16 luma prediction modes described in Section 6.4.6.2 and illustrated in Figure 6.26, except that the numbering of the modes is different. The modes are DC (mode 0), horizontal (mode 1), vertical (mode 2) and plane (mode 3).

6.4.6.4 Signalling Intra Prediction Modes

The choice of intra prediction mode for each 4 × 4 block must be signalled to the decoder and this could potentially require a large number of bits. However, intra modes for neighbouring 4 × 4 blocks are often correlated. For example, let A, B and E be the left, upper and current 4 × 4 blocks respectively (the same labelling as Figure 6.18). If previously-encoded 4 × 4 blocks A and B are predicted using mode 1, it is probable that the best mode for block E (current block) is also mode 1. To take advantage of this correlation, predictive coding is used to signal 4 × 4 intra modes.

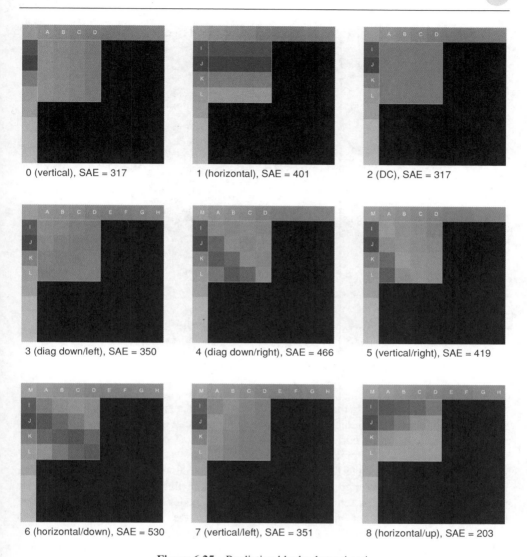

0 (vertical), SAE = 317

1 (horizontal), SAE = 401

2 (DC), SAE = 317

3 (diag down/left), SAE = 350

4 (diag down/right), SAE = 466

5 (vertical/right), SAE = 419

6 (horizontal/down), SAE = 530

7 (vertical/left), SAE = 351

8 (horizontal/up), SAE = 203

Figure 6.25   Prediction blocks, luma 4 × 4

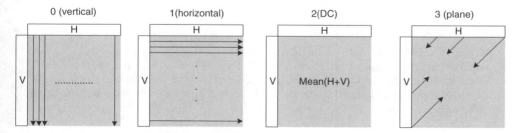

Figure 6.26   Intra 16 × 16 prediction modes

16x16 luminance block to be predicted

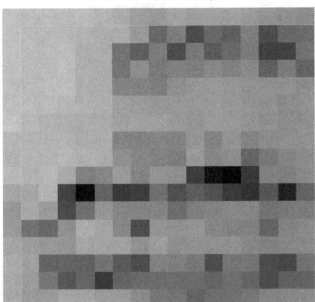

Figure 6.27   16 × 16 macroblock

For each current block E, the encoder and decoder calculate the most probable prediction mode, the minimum of the prediction modes of A and B. If either of these neighbouring blocks is not available (outside the current slice or not coded in Intra4×4 mode), the corresponding value A or B is set to 2 (DC prediction mode).

The encoder sends a flag for each 4 × 4 block, prev_intra4×4_pred_mode. If the flag is '1', the most probable prediction mode is used. If the flag is '0', another parameter rem_intra4×4_pred_mode is sent to indicate a change of mode. If rem_intra4×4_pred_mode is smaller than the current most probable mode then the prediction mode is set to rem_intra4×4_pred_mode, otherwise the prediction mode is set to (rem_intra4×4_pred_mode +1). In this way, only eight values of rem_intra4×4_pred_mode are required (0 to 7) to signal the current intra mode (0 to 8).

Example

Blocks A and B have been predicted using mode 3 (diagonal down-left) and 1 (horizontal) respectively. The most probable mode for block E is therefore 1 (horizontal). prev_intra4×4_pred_mode is set to '0' and so rem_intra4×4_pred_mode is sent. Depending on the value of rem_intra4×4_pred_mode, one of the eight remaining prediction modes (listed in Table 6.4) may be chosen.

The prediction mode for luma coded in Intra-16×16 mode or chroma coded in Intra mode is signalled in the macroblock header and predictive coding of the mode is not used in these cases.

Table 6.4 Choice of prediction mode (most probable mode = 1)

rem_intra4×4_pred_mode	prediction mode for block C
0	0
1	2
2	3
3	4
4	5
5	6
6	7
7	8

0 (vertical), SAE = 3985

1 (horizontal), SAE = 5097

2 (DC), SAE = 4991

3 (plane), SAE = 2539

Figure 6.28 Prediction blocks, intra 16×16

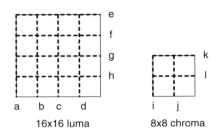

Figure 6.29   Edge filtering order in a macroblock

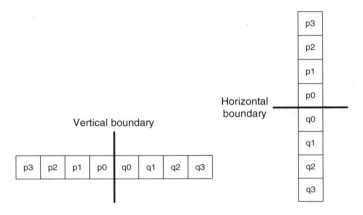

Figure 6.30   Samples adjacent to vertical and horizontal boundaries

6.4.7 Deblocking Filter

A filter is applied to each decoded macroblock to reduce blocking distortion. The deblocking filter is applied after the inverse transform in the encoder (before reconstructing and storing the macroblock for future predictions) and in the decoder (before reconstructing and display-ing the macroblock). The filter smooths block edges, improving the appearance of decoded frames. The filtered image is used for motion-compensated prediction of future frames and this can improve compression performance because the filtered image is often a more faithful reproduction of the original frame than a blocky, unfiltered image[3]. The default operation of the filter is as follows; it is possible for the encoder to alter the filter strength or to disable the filter.

Filtering is applied to vertical or horizontal edges of 4 × 4 blocks in a macroblock (except for edges on slice boundaries), in the following order.

1. Filter 4 vertical boundaries of the luma component (in order a, b, c, d in Figure 6.29).
2. Filter 4 horizontal boundaries of the luma component (in order e, f, g, h, Figure 6.29).
3. Filter 2 vertical boundaries of each chroma component (i, j).
4. Filter 2 horizontal boundaries of each chroma component (k, l).

Each filtering operation affects up to *three* samples on either side of the boundary. Figure 6.30 shows four samples on either side of a vertical or horizontal boundary in adjacent blocks

---

[3] Intra-coded macroblocks are filtered, but intra prediction (Section 6.4.6) is carried out using *unfiltered* reconstructed macroblocks to form the prediction.

p and q (p0, p1, p2, p3 and q0, q1, q2, q3). The 'strength' of the filter (the amount of filtering) depends on the current quantiser, the coding modes of neighbouring blocks and the gradient of image samples across the boundary.

Boundary Strength

The choice of filtering outcome depends on the *boundary strength* and on the *gradient* of image samples across the boundary. The boundary strength parameter bS is chosen according to the following rules (for coding of progressive frames):

p and/or q is intra coded *and* boundary is a macroblock boundary	$bS = 4$ (strongest filtering)
p and q are intra coded and boundary is *not* a macroblock boundary	$bS = 3$
neither p or q is intra coded; p and q contain coded coefficients	$bS = 2$
neither p or q is intra coded; neither p or q contain coded coefficients; p and q use different reference pictures *or* a different number of reference pictures or have motion vector values that differ by one luma sample or more	$bS = 1$
otherwise	$bS = 0$ (no filtering)

The result of applying these rules is that the filter is stronger at places where there is likely to be significant blocking distortion, such as the boundary of an intra coded macroblock or a boundary between blocks that contain coded coefficients.

Filter Decision

A group of samples from the set (p2, p1, p0, q0, q1, q2) is filtered only if:

(a) BS > 0 and
(b) $|p0-q0| < \alpha$ *and* $|p1-p0| < \beta$ *and* $|q1-q0| \leq \beta$.

α and β are thresholds defined in the standard; they increase with the average quantiser parameter QP of the two blocks p and q. The effect of the filter decision is to 'switch off' the filter when there is a significant change (gradient) across the block boundary in the original image. When QP is small, anything other than a very small gradient across the boundary is likely to be due to image features (rather than blocking effects) that should be preserved and so the thresholds α and β are low. When QP is larger, blocking distortion is likely to be more significant and α, β are higher so that more boundary samples are filtered.

Example

Figure 6.31 shows the 16 × 16 luma component of a macroblock (without any blocking distortion) with four 4 × 4 blocks a, b, c and d highlighted. Assuming a medium to large value of QP, the block boundary between a and b is likely to be filtered because the gradient across this boundary is small. There are no significant image features to preserve and blocking distortion will be obvious on this boundary. However, there is a significant change in luminance across the boundary between c and d due to a horizontal image feature and so the filter is switched off to preserve this strong feature.

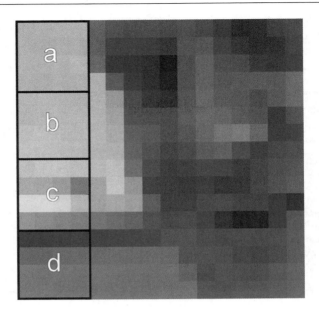

Figure 6.31   16 × 16 luma macroblock showing block edges

Filter Implementation

(a) bS ∈ {1,2,3}

A 4-tap filter is applied with inputs p1, p0, q0 and q1, producing filtered outputs p'0 and q'0. If |p2 − p0| is less than threshold β, another four-tap filter is applied with inputs p2, p1, p0 and q0, producing filtered output p'1 (luma only). If |q2 − q0| is less than the threshold β, a four-tap filter is applied with inputs q2, q1, q0 and p0, producing filtered output q'1 (luma only).

(b) bS = 4

If |p2 − p0| < β and |p0 − q0| < round (α/4) and this is a luma block:

     p'0 is produced by five-tap filtering of p2, p1, p0, q0 and q1,

     p'1 is produced by four-tap filtering of p2, p1, p0 and q0,

     p'2 is produced by five-tap filtering of p3, p2, p1, p0 and q0,

else:

     p'0 is produced by three-tap filtering of p1, p0 and q1.

If |q2 − q0| < β and |p0 − q0| < round(α/4) and this is a luma block:

     q'0 is produced by five-tap filtering of q2, q1, q0, p0 and p1,

     q'1 is produced by four-tap filtering of q2, q1, q0 and p0,

     q'2 is produced by five-tap filtering of q3, q2, q1, q0 and p0,

else:

     q'0 is produced by three-tap filtering of q1, q0 and p1.

Example

A video clip is encoded with a fixed Quantisation Parameter of 36 (relatively high quantisation). Figure 6.32 shows an original frame from the clip and Figure 6.33 shows the same frame after inter coding and decoding, with the loop filter disabled. Note the obvious blocking artefacts and note also the effect of varying motion-compensation block sizes (for example, 16×16 blocks in the background to the left of the picture, 4×4 blocks around the arm). With the loop filter enabled (Figure 6.34) there is still some obvious distortion but most of the block edges have disappeared or faded. Note that sharp contrast boundaries (such as the line of the arm against the dark piano) are preserved by the filter whilst block edges in smoother regions of the picture (such as the background to the left) are smoothed. In this example the loop filter makes only a small contribution to compression efficiency: the encoded bitrate is around 1.5% smaller and the PSNR around 1% larger for the sequence with the filter. However, the subjective quality of the filtered sequence is significantly better. The coding performance gain provided by the filter depends on the bitrate and sequence content.

Figure 6.35 and Figure 6.36 show the un-filtered and filtered frame respectively, this time with a lower quantiser parameter ($QP = 32$).

6.4.8 Transform and Quantisation

H.264 uses three transforms depending on the type of residual data that is to be coded: a Hadamard transform for the 4×4 array of luma DC coefficients in intra macroblocks predicted in 16×16 mode, a Hadamard transform for the 2×2 array of chroma DC coefficients (in any macroblock) and a DCT-based transform for all other 4×4 blocks in the residual data.

Data within a macroblock are transmitted in the order shown in Figure 6.47. If the macroblock is coded in 16×16 Intra mode, then the block labelled '-1', containing the transformed DC coefficient of each 4×4 luma block, is transmitted first. Next, the luma residual blocks 0–15 are transmitted in the order shown (the DC coefficients in a macroblock coded in 16×16 Intra mode are not sent). Blocks 16 and 17 are sent, containing a 2×2 array of DC coefficients from the Cb and Cr chroma components respectively and finally, chroma residual blocks 18–25 (without DC coefficients) are sent.

6.4.8.1 4×4 Residual Transform and Quantisation (blocks 0–15, 18–25)

This transform operates on 4×4 blocks of residual data (labelled 0–15 and 18–25 in Figure 6.37) after motion-compensated prediction or Intra prediction. The H.264 transform [3] is based on the DCT but with some fundamental differences:

1. It is an integer transform (all operations can be carried out using integer arithmetic, without loss of decoding accuracy).
2. It is possible to ensure zero mismatch between encoder and decoder inverse transforms (using integer arithmetic).
3. The core part of the transform can be implemented using only additions and shifts.
4. A scaling multiplication (part of the transform) is integrated into the quantiser, reducing the total number of multiplications.

The inverse quantisation (scaling) and inverse transform operations can be carried out using 16-bit integer arithmetic (footnote: except in the case of certain anomalous residual data patterns) with only a single multiply per coefficient, without any loss of accuracy.

Figure 6.32   Original frame (violin frame 2)

Figure 6.33   Reconstructed, QP = 36 (no filter)   **Figure 6.34**   Reconstructed, QP = 36 (with filter

Figure 6.35   Reconstructed, QP = 32 (no filter)   **Figure 6.36**   Reconstructed, QP = 32 (with filter

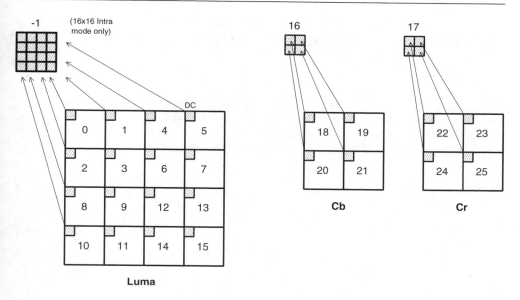

Figure 6.37 Scanning order of residual blocks within a macroblock

Development from the 4 × 4 DCT

Recall from Chapter 3 that the 4 × 4 DCT is given by:

$$\mathbf{Y} = \mathbf{A}\mathbf{X}\mathbf{A}^{\mathrm{T}} = \begin{bmatrix} a & a & a & a \\ b & c & -c & -b \\ a & -a & -a & a \\ c & -b & b & -c \end{bmatrix} \begin{bmatrix} \mathbf{X} \end{bmatrix} \begin{bmatrix} a & b & a & c \\ a & c & -a & -b \\ a & -c & -a & b \\ a & -b & a & -c \end{bmatrix} \qquad (6.1)$$

where:

$$a = \frac{1}{2}, \qquad b = \sqrt{\frac{1}{2}}\cos\left(\frac{\pi}{8}\right), \qquad c = \sqrt{\frac{1}{2}}\cos\left(\frac{3\pi}{8}\right)$$

This matrix multiplication can be factorised [3] to the following equivalent form (equation 6.2):

$$\mathbf{Y} = (CXC^{\mathrm{T}}) \otimes \mathbf{E} = \left(\begin{bmatrix} 1 & 1 & 1 & 1 \\ 1 & d & -d & -1 \\ 1 & -1 & -1 & 1 \\ d & -1 & 1 & -d \end{bmatrix} \begin{bmatrix} \mathbf{X} \end{bmatrix} \begin{bmatrix} 1 & 1 & 1 & d \\ 1 & d & -1 & -1 \\ 1 & -d & -1 & 1 \\ 1 & -1 & 1 & -d \end{bmatrix} \right) \otimes \begin{bmatrix} a^2 & ab & a^2 & ab \\ ab & b^2 & ab & b^2 \\ a^2 & ab & a^2 & ab \\ ab & b^2 & ab & b^2 \end{bmatrix}$$

$$(6.2)$$

$\mathbf{CXC}^{\mathrm{T}}$ is a 'core' 2D transform. $\mathbf{E}$ is a matrix of scaling factors and the symbol $\otimes$ indicates that each element of $(\mathbf{CXC}^{\mathrm{T}})$ is multiplied by the scaling factor in the same position in matrix $\mathbf{E}$ (scalar multiplication rather than matrix multiplication). The constants a and b are as before and d is c/b (approximately 0.414).

To simplify the implementation of the transform, d is approximated by 0.5. In order to ensure that the transform remains orthogonal, b also needs to be modified so that:

$$a = \frac{1}{2}, \qquad b = \sqrt{\frac{2}{5}}, \qquad d = \frac{1}{2}$$

The 2nd and 4th rows of matrix $\mathbf{C}$ and the 2nd and 4th columns of matrix $\mathbf{C}^\mathrm{T}$ are scaled by a factor of two and the post-scaling matrix $\mathbf{E}$ is scaled down to compensate, avoiding multiplications by half in the 'core' transform $\mathbf{CXC}^\mathrm{T}$ which could result in loss of accuracy using integer arithmetic. The final forward transform becomes:

$$Y = C_f X C_f^\mathrm{T} \otimes E_f = \left(\begin{bmatrix} 1 & 1 & 1 & 1 \\ 2 & 1 & -1 & -2 \\ 1 & -1 & -1 & 1 \\ 1 & -2 & 2 & -1 \end{bmatrix} \begin{bmatrix} & & & \\ & \mathbf{X} & & \\ & & & \end{bmatrix} \begin{bmatrix} 1 & 2 & 1 & 1 \\ 1 & 1 & -1 & -2 \\ 1 & -1 & -1 & 2 \\ 1 & -2 & 1 & -1 \end{bmatrix} \right) \otimes \begin{bmatrix} a^2 & \frac{ab}{2} & a^2 & \frac{ab}{2} \\ \frac{ab}{2} & \frac{b^2}{4} & \frac{ab}{2} & \frac{b^2}{4} \\ a^2 & \frac{ab}{2} & a^2 & \frac{ab}{2} \\ \frac{ab}{2} & \frac{b^2}{4} & \frac{ab}{2} & \frac{b^2}{4} \end{bmatrix}$$

$$(6.3)$$

This transform is an approximation to the 4×4 DCT but because of the change to factors d and b, the result of the new transform will not be identical to the 4×4 DCT.

Example

Compare the output of the 4×4 approximate transform with the output of a 'true' 4×4 DCT, for input block $\mathbf{X}$:

$\mathbf{X}$:

	$j=0$	1	2	3
$i=0$	5	11	8	10
1	9	8	4	12
2	1	10	11	4
3	19	6	15	7

DCT output:

$$Y = \mathbf{AXA}^\mathrm{T} = \begin{bmatrix} 35.0 & -0.079 & -1.5 & 1.115 \\ -3.299 & -4.768 & 0.443 & -9.010 \\ 5.5 & 3.029 & 2.0 & 4.699 \\ -4.045 & -3.010 & -9.384 & -1.232 \end{bmatrix}$$

Approximate transform output:

$$Y' = (\mathbf{CXC}^\mathrm{T}) \otimes \mathbf{E}_f = \begin{bmatrix} 35.0 & -0.158 & -1.5 & 1.107 \\ -3.004 & -3.900 & 1.107 & -9.200 \\ 5.5 & 2.688 & 2.0 & 4.901 \\ -4.269 & -3.200 & -9.329 & -2.100 \end{bmatrix}$$

Difference between DCT and integer transform outputs:

$$
\mathbf{Y} - \mathbf{Y}' = \begin{bmatrix} 0 & 0.079 & 0 & 0.008 \\ -0.295 & -0.868 & -0.664 & 0.190 \\ 0 & 0.341 & 0 & -0.203 \\ 0.224 & 0.190 & -0.055 & 0.868 \end{bmatrix}
$$

There is clearly a difference in the output coefficients that depend on b or d. In the context of the H.264 CODEC, the approximate transform has almost identical compression performance to the DCT and has a number of important advantages. The 'core' part of the transform, $\mathbf{CXC^T}$, can be carried out with integer arithmetic using only additions, subtractions and shifts. The dynamic range of the transform operations is such that 16-bit arithmetic may be used throughout (except in the case of certain anomalous input patterns) since the inputs are in the range ± 255. The post-scaling operation $\otimes \mathbf{E}_f$ requires one multiplication for every coefficient which can be 'absorbed' into the quantisation process (see below).

The inverse transform is given by Equation 6.4. The H.264 standard [1] defines this transform explicitly as a sequence of arithmetic operations:

$$
\mathbf{Y} = \mathbf{C}_i^T(\mathbf{Y} \otimes \mathbf{E}_i)\mathbf{C}_i = \begin{bmatrix} 1 & 1 & 1 & \frac{1}{2} \\ 1 & \frac{1}{2} & -1 & -1 \\ 1 & -\frac{1}{2} & -1 & 1 \\ 1 & -1 & 1 & -\frac{1}{2} \end{bmatrix} \left(\begin{bmatrix} \mathbf{X} \end{bmatrix} \otimes \begin{bmatrix} a^2 & ab & a^2 & ab \\ ab & b^2 & ab & b^2 \\ a^2 & ab & a^2 & ab \\ ab & b^2 & ab & b^2 \end{bmatrix} \right) \begin{bmatrix} 1 & 1 & 1 & 1 \\ 1 & \frac{1}{2} & -\frac{1}{2} & -1 \\ 1 & -1 & -1 & 1 \\ \frac{1}{2} & -1 & 1 & -\frac{1}{2} \end{bmatrix}
$$

$$(6.4)$$

This time, $\mathbf{Y}$ is *pre-scaled* by multiplying each coefficient by the appropriate weighting factor from matrix $\mathbf{E}_i$. Note the factors $\pm 1/2$ in the matrices $\mathbf{C}$ and $\mathbf{C^T}$ which can be implemented by a right-shift without a significant loss of accuracy because the coefficients $\mathbf{Y}$ are pre-scaled.

The forward and inverse transforms are orthogonal, i.e. $\mathbf{T^{-1}(T(X))} = \mathbf{X}$.

Quantisation

H.264 assumes a scalar quantiser (see Chapter 3). The mechanisms of the forward and inverse quantisers are complicated by the requirements to (a) avoid division and/or floating point arithmetic and (b) incorporate the post- and pre-scaling matrices $\mathbf{E}_f$ and $\mathbf{E}_i$ described above.

The basic forward quantiser [3] operation is:

$$Z_{ij} = round(Y_{ij}/Qstep)$$

where Y_{ij} is a coefficient of the transform described above, Qstep is a quantizer step size and Z_{ij} is a quantised coefficient. The rounding operation here (and throughout this section) need not round to the nearest integer; for example, biasing the 'round' operation towards smaller integers can give perceptual quality improvements.

A total of 52 values of Qstep are supported by the standard, indexed by a Quantisation Parameter, QP (Table 6.5). Qstep doubles in size for every increment of 6 in QP. The wide range of quantiser step sizes makes it possible for an encoder to control the tradeoff accurately and flexibly between bit rate and quality. The values of QP can be different for luma and chroma. Both parameters are in the range 0–51 and the default is that the chroma parameter

Table 6.5   Quantisation step sizes in H.264 CODEC

QP	0	1	2	3	4	5	6	7	8	9	10	11	12	...
QStep	0.625	0.6875	0.8125	0.875	1	1.125	1.25	1.375	1.625	1.75	2	2.25	2.5	...
QP	...	18	...	24	...	30	...	36	...	42	...	48	...	51
QStep		5		10		20		40		80		160		224

QP_C is derived from QP_Y so that QP_C is less that QP_Y for $QP_Y > 30$. A user-defined mapping between QP_Y and QP_C may be signalled in a Picture Parameter Set.

The post-scaling factor a^2, $ab/2$ or $b^2/4$ (equation 6.3) is incorporated into the forward quantiser. First, the input block **X** is transformed to give a block of unscaled coefficients $\mathbf{W} = \mathbf{CXC}^T$. Then, each coefficient W_{ij} is quantised and scaled in a single operation:

$$Z_{ij} = round\left(W_{ij}.\frac{PF}{Qstep}\right) \tag{6.5}$$

PF is a^2, $ab/2$ or $b^2/4$ depending on the position (i, j) (see equation 6.3):

Position	PF
(0,0), (2,0), (0,2) or (2,2)	a^2
(1,1), (1,3), (3,1) or (3,3)	$b^2/4$
other	$ab/2$

In order to simplify the arithmetic), the factor $(PF/Qstep)$ is implemented in the reference model software [4] as a multiplication by a factor MF and a right-shift, avoiding any division operations:

$$Z_{ij} = round\left(W_{ij}.\frac{MF}{2^{qbits}}\right)$$

where

$$\frac{MF}{2^{qbits}} = \frac{PF}{Qstep}$$

and

$$qbits = 15 + floor(QP/6) \tag{6.6}$$

In integer arithmetic, Equation 6.6 can be implemented as follows:

$$|Z_{ij}| = (|W_{ij}|.MF + f) >> qbits$$
$$sign(Z_{ij}) = sign(W_{ij}) \tag{6.7}$$

where $>>$ indicates a binary shift right. In the reference model software, f is $2^{qbits}/3$ for Intra blocks or $2^{qbits}/6$ for Inter blocks.

Table 6.6 Multiplication factor MF

QP	Positions (0,0),(2,0),(2,2),(0,2)	Positions (1,1),(1,3),(3,1),(3,3)	Other positions
0	13107	5243	8066
1	11916	4660	7490
2	10082	4194	6554
3	9362	3647	5825
4	8192	3355	5243
5	7282	2893	4559

Example

$QP = 4$ and $(i,j) = (0,0)$.

$Qstep = 1.0$, $PF = a^2 = 0.25$ and $qbits = 15$, hence $2^{qbits} = 32768$.

$$\frac{MF}{2^{qbits}} = \frac{PF}{Qstep}, \qquad MF = (32768 \times 0.25)/1 = 8192$$

The first six values of MF (for each coefficient position) used by the H.264 reference software encoder are given in Table 6.6. The 2nd and 3rd columns of this table (positions with factors $b^2/4$ and $ab/2$) have been modified slightly[4] from the results of equation 6.6.
For $QP > 5$, the factors MF remain unchanged but the divisor 2^{qbits} increases by a factor of two for each increment of six in QP. For example, $qbits = 16$ for $6 \leq QP \leq 11$, $qbits = 17$ for $12 \leq QP \leq 17$ and so on.

ReScaling
The basic scaling (or 'inverse quantiser') operation is:

$$Y'_{ij} = Z_{ij} Qstep \tag{6.8}$$

The pre-scaling factor for the inverse transform (from matrix $\mathbf{E}_i$, containing values a^2, ab and b^2 depending on the coefficient position) is incorporated in this operation, together with a constant scaling factor of 64 to avoid rounding errors:

$$W'_{ij} = Z_{ij} Qstep \cdot PF \cdot 64 \tag{6.9}$$

W'_{ij} is a scaled coefficient which is transformed by the core inverse transform $\mathbf{C}_i^T \mathbf{W} \mathbf{C}_i$ (Equation 6.4). The values at the output of the inverse transform are divided by 64 to remove the scaling factor (this can be implemented using only an addition and a right-shift). The H.264 standard does not specify Qstep or PF directly. Instead, the parameter $V = (Qstep.PF.64)$ is defined for $0 \leq QP \leq 5$ and for each coefficient position so that the scaling

---

[4] It is acceptable to modify a forward quantiser, for example in order to improve perceptual quality at the decoder, since only the rescaling (inverse quantiser) process is standardised.

operation becomes:

$$W'_{ij} = Z_{ij} V_{ij} \cdot 2^{floor(QP/6)}$$ (6.10)

Example

$QP = 3$ and $(i, j) = (1, 2)$

$Qstep = 0.875$ and $2^{floor(QP/6)} = 1$

$PF = ab = 0.3162$

$V = (Qstep \cdot PF \cdot 64) = 0.875 \times 0.3162 \times 65 \cong 18$

$W'_{ij} = Z_{ij} \times 18 \times 1$

The values of V defined in the standard for $0 \le QP \le 5$ are shown in Table 6.7.

The factor $2^{floor(QP/6)}$ in Equation 6.10 causes the sclaed output increase by a factor of two for every increment of six in QP.

6.4.9 4 × 4 Luma DC Coefficient Transform and Quantisation (16 × 16 Intra-mode Only)

If the macroblock is encoded in 16×16 Intra prediction mode (i.e. the entire 16×16 luma component is predicted from neighbouring samples), each 4×4 residual block is first transformed using the 'core' transform described above ($\mathbf{C}_f \mathbf{X} \mathbf{C}_f^T$). The DC coefficient of each 4×4 block is then transformed again using a 4×4 Hadamard transform:

$$\mathbf{Y}_D = \left(\begin{bmatrix} 1 & 1 & 1 & 1 \\ 1 & 1 & -1 & -1 \\ 1 & -1 & -1 & 1 \\ 1 & -1 & 1 & -1 \end{bmatrix} \begin{bmatrix} & & \\ & \mathbf{W}_D & \\ & & \end{bmatrix} \begin{bmatrix} 1 & 1 & 1 & 1 \\ 1 & 1 & -1 & -1 \\ 1 & -1 & -1 & 1 \\ 1 & -1 & 1 & -1 \end{bmatrix} \right) /2$$ (6.11)

$\mathbf{W}_D$ is the block of 4×4 DC coefficients and $\mathbf{Y}_D$ is the block after transformation. The output coefficients $Y_{D(i,j)}$ are quantised to produce a block of quantised DC coefficients:

$$|Z_{D(i,j)}| = (|Y_{D(i,j)}| MF_{(0,0)} + 2f) >> (qbits + 1)$$
$$\text{sign}(Z_{D(i,j)}) = \text{sign}(Y_{D(i,j)})$$ (6.12)

$MF_{(0,0)}$ is the multiplication factor for position (0,0) in Table 6.6 and f, $qbits$ are defined as before.

At the decoder, an inverse Hadamard transform is applied *followed by* rescaling (note that the order is not reversed as might be expected):

$$\mathbf{W}_{QD} = \left(\begin{bmatrix} 1 & 1 & 1 & 1 \\ 1 & 1 & -1 & -1 \\ 1 & -1 & -1 & 1 \\ 1 & -1 & 1 & -1 \end{bmatrix} \begin{bmatrix} & & \\ & \mathbf{Z}_D & \\ & & \end{bmatrix} \begin{bmatrix} 1 & 1 & 1 & 1 \\ 1 & 1 & -1 & -1 \\ 1 & -1 & -1 & 1 \\ 1 & -1 & 1 & -1 \end{bmatrix} \right)$$ (6.13)

Table 6.7 Scaling factor V

QP	Positions (0,0),(2,0),(2,2),(0,2)	Positions (1,1),(1,3),(3,1),(3,3)	Other positions
0	10	16	13
1	11	18	14
2	13	20	16
3	14	23	18
4	16	25	20
5	18	29	23

Decoder scaling is performed by:

$$W'_{D(i,j)} = W_{QD(i,j)} V_{(0,0)} 2^{floor}(QP/6) - 2 \qquad (QP \geq 12)$$
$$W'_{D(i,j)} = \left[W_{QD(i,j)} V_{(0,0)} + 2^{1-floor(QP/6)}\right] >> (2 - floor(QP/6)) \qquad (QP < 12)$$

$$(6.14)$$

$V_{(0,0)}$ is the scaling factor V for position (0,0) in Table 6.7. Because $V_{(0,0)}$ is constant throughout the block, rescaling and inverse transformation can be applied in any order. The specified order (inverse transform first, then scaling) is designed to maximise the dynamic range of the inverse transform.

The rescaled DC coefficients $\mathbf{W}'_D$ are inserted into their respective 4×4 blocks and each 4×4 block of coefficients is inverse transformed using the core DCT-based inverse transform $(\mathbf{C}_i^T \mathbf{W}' \mathbf{C}_i)$. In a 16×16 intra-coded macroblock, much of the energy is concentrated in the DC coefficients of each 4×4 block which tend to be highly correlated. After this extra transform, the energy is concentrated further into a small number of significant coefficients.

6.4.10 2 × 2 Chroma DC Coefficient Transform and Quantisation

Each 4×4 block in the chroma components is transformed as described in Section 6.4.8.1. The DC coefficients of each 4×4 block of chroma coefficients are grouped in a 2×2 block ($\mathbf{W}_D$) and are further transformed prior to quantisation:

$$\mathbf{W}_{QD} = \begin{bmatrix} 1 & 1 \\ 1 & -1 \end{bmatrix} \begin{bmatrix} \mathbf{W}_D \end{bmatrix} \begin{bmatrix} 1 & 1 \\ 1 & -1 \end{bmatrix} \qquad (6.15)$$

Quantisation of the 2×2 output block $\mathbf{Y}_D$ is performed by:

$$|Z_{D(i,j)}| = (|Y_{D(i,j)}| . MF_{(0,0)} + 2f) >> (qbits + 1) \qquad (6.16)$$
$$\text{sign}(Z_{D(i,j)}) = \text{sign}(Y_{D(i,j)})$$

$MF_{(0,0)}$ is the multiplication factor for position (0,0) in Table 6.6, f and $qbits$ are defined as before.

During decoding, the inverse transform is applied before scaling:

$$\mathbf{W}_{QD} = \begin{bmatrix} 1 & 1 \\ 1 & -1 \end{bmatrix} \begin{bmatrix} \mathbf{Z}_D \end{bmatrix} \begin{bmatrix} 1 & 1 \\ 1 & -1 \end{bmatrix} \qquad (6.17)$$

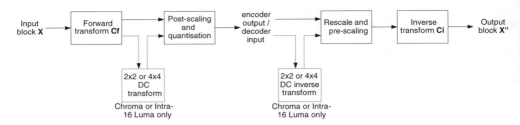

Figure 6.38   Transform, quantisation, rescale and inverse transform flow diagram

Scaling is performed by:

$$W'_{D(i,j)} = W_{QD(i,j)}.V_{(0,0)}.2^{floor(QP/6)-1} \quad \text{(if } QP \geq 6)$$

$$W'_{D(i,j)} = \left[W_{QD(i,j)}.V_{(0,0)} \right] >> 1 \quad \text{(if } QP < 6)$$

The rescaled coefficients are replaced in their respective 4×4 blocks of chroma coefficients which are then transformed as above ($\mathbf{C}_i^T \mathbf{W}' \mathbf{C}_i$). As with the Intra luma DC coefficients, the extra transform helps to de-correlate the 2×2 chroma DC coefficients and improves compression performance.

6.4.11 The Complete Transform, Quantisation, Rescaling and Inverse Transform Process

The complete process from input residual block $\mathbf{X}$ to output residual block $\mathbf{X}'$ is described below and illustrated in Figure 6.38.

Encoding:
1. Input: 4×4 residual samples: $\mathbf{X}$
2. Forward 'core' transform: $\mathbf{W} = \mathbf{C}_f \mathbf{X} \mathbf{C}_f^T$
   (followed by forward transform for Chroma DC or Intra-16 Luma DC coefficients).
3. Post-scaling and quantisation: $\mathbf{Z} = \mathbf{W}.\text{round}(PF/Qstep)$
   (different for Chroma DC or Intra-16 Luma DC).

Decoding:
   (Inverse transform for Chroma DC or Intra-16 Luma DC coefficients)
4. Decoder scaling (incorporating inverse transform pre-scaling): $\mathbf{W}' = \mathbf{Z}.Qstep.PF.64$
   (different for Chroma DC or Intra-16 Luma DC).
5. Inverse 'core' transform: $\mathbf{X}' = \mathbf{C}_i^T \mathbf{W}' \mathbf{C}_i$
6. Post-scaling: $\mathbf{X}'' = \text{round}(X'/64)$
7. Output: 4×4 residual samples: $\mathbf{X}''$

Example (luma 4 × 4 residual block, Intra mode)

$QP = 10$

Input block **X**:

	$j = 0$	1	2	3
$i = 0$	5	11	8	10
1	9	8	4	12
2	1	10	11	4
3	19	6	15	7

Output of 'core' transform **W**:

	$j = 0$	1	2	3
$i = 0$	140	-1	-6	7
1	-19	-39	7	-92
2	22	17	8	31
3	-27	-32	-59	-21

$MF = 8192$, 3355 or 5243 (depending on the coefficient position), $qbits = 16$ and f is $2^{qbits}/3$. Output of forward quantizer **Z**:

	$j = 0$	1	2	3
$i = 0$	17	0	-1	0
1	-1	-2	0	-5
2	3	1	1	2
3	-2	-1	-5	-1

$V = 16$, 25 or 20 (depending on position) and $2^{\text{floor}}(QP/6) = 2^1 = 2$. Output of rescale **W'**:

	$j = 0$	1	2	3
$i = 0$	544	0	-32	0
1	-40	-100	0	-250
2	96	40	32	80
3	-80	-50	-200	-50

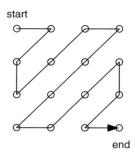

start

end

Figure 6.39   Zig-zag scan for 4 × 4 luma block (frame mode)

Output of 'core' inverse transform $\mathbf{X}''$ (after division by 64 and rounding):

	$j = 0$	1	2	3
$i = 0$	4	13	8	10
1	8	8	4	12
2	1	10	10	3
3	18	5	14	7

6.4.12 Reordering

In the encoder, each 4 × 4 block of quantised transform coefficients is mapped to a 16-element array in a zig-zag order (Figure 6.39). In a macroblock encoded in 16 × 16 Intra mode, the DC coefficients (top-left) of each 4 × 4 luminance block are scanned first and these DC coefficients form a 4 × 4 array that is scanned in the order of Figure 6.39. This leaves 15 AC coefficients in each luma block that are scanned starting from the 2nd position in Figure 6.39. Similarly, the 2 × 2 DC coefficients of each chroma component are first scanned (in raster order) and then the 15 AC coefficients in each chroma 4 × 4 block are scanned starting from the 2nd position.

6.4.13 Entropy Coding

Above the slice layer, syntax elements are encoded as fixed- or variable-length binary codes. At the slice layer and below, elements are coded using either variable-length codes (VLCs) or context-adaptive arithmetic coding (CABAC) depending on the entropy encoding mode. When entropy_coding_mode is set to 0, residual block data is coded using a context-adaptive variable length coding (CAVLC) scheme and other variable-length coded units are coded using Exp-Golomb codes. Parameters that require to be encoded and transmitted include the following (Table 6.8).

Table 6.8 Examples of parameters to be encoded

Parameters	Description
Sequence-, picture- and slice-layer syntax elements	Headers and parameters
Macroblock type mb_type	Prediction method for each coded macroblock
Coded block pattern	Indicates which blocks within a macroblock contain coded coefficients
Quantiser parameter	Transmitted as a delta value from the previous value of QP
Reference frame index	Identify reference frame(s) for inter prediction
Motion vector	Transmitted as a difference (mvd) from predicted motion vector
Residual data	Coefficient data for each 4×4 or 2×2 block

Table 6.9 Exp-Golomb codewords

code_num	Codeword
0	1
1	010
2	011
3	00100
4	00101
5	00110
6	00111
7	0001000
8	0001001
...	...

6.4.13.1 Exp-Golomb Entropy Coding

Exp-Golomb codes (Exponential Golomb codes, [5]) are variable length codes with a regular construction. It is clear from examining the first few codewords (Table 6.9) that they are constructed in a logical way:

$$[\text{M zeros}][1][\text{INFO}]$$

INFO is an M-bit field carrying information. The first codeword has no leading zero or trailing INFO. Codewords 1 and 2 have a single-bit INFO field, codewords 3–6 have a two-bit INFO field and so on. The length of each Exp-Golomb codeword is $(2M + 1)$ bits and each codeword can be constructed by the encoder based on its index *code_num*:

$$M = \text{floor}(\log_2[\text{code\_num} + 1])$$
$$\text{INFO} = \text{code\_num} + 1 - 2^M$$

A codeword can be decoded as follows:

1. Read in M leading zeros followed by 1.
2. Read M-bit INFO field.
3. code_num $= 2^M +$ INFO $- 1$

(For codeword 0, INFO and M are zero.)

A parameter *k* to be encoded is mapped to code_num in one of the following ways:

Mapping type	Description
ue	Unsigned direct mapping, code_num = k. Used for macroblock type, reference frame index and others.
te	A version of the Exp-Golomb codeword table in which short codewords are truncated.
se	Signed mapping, used for motion vector difference, delta QP and others. k is mapped to code_num as follows (Table 6.10). code_num = 2|k| (k ≤ 0) code_num = 2|k|− 1 (k > 0)
me	Mapped symbols, parameter k is mapped to code_num according to a table specified in the standard. Table 6.11 lists a small part of the coded_block_pattern table for Inter predicted macroblocks, indicating which 8 × 8 blocks in a macroblock contain nonzero coefficients.

Table 6.10   Signed mapping se

k	code_num
0	0
1	1
−1	2
2	3
−2	4
3	5
. . .	. . .

Table 6.11   Part of coded_block_pattern table

coded_block_pattern (Inter prediction)	code_num
0 (no nonzero blocks)	0
16 (chroma DC block nonzero)	1
1 (top-left 8 × 8 luma block nonzero)	2
2 (top-right 8 × 8 luma block nonzero)	3
4 (lower-left 8 × 8 luma block nonzero)	4
8 (lower-right 8 × 8 luma block nonzero)	5
32 (chroma DC and AC blocks nonzero)	6
3 (top-left and top-right 8 × 8 luma blocks nonzero)	7
. . .	. . .

Each of these mappings (ue, te, se and me) is designed to produce short codewords for frequently-occurring values and longer codewords for less common parameter values. For example, inter macroblock type P_L0_16 × 16 (prediction of 16 × 16 luma partition from a previous picture) is assigned code_num 0 because it occurs frequently; macroblock type P_8 × 8 (prediction of 8 × 8 luma partition from a previous picture) is assigned code_num 3 because it occurs less frequently; the commonly-occurring motion vector difference (MVD) value of 0 maps to code_num 0 whereas the less-common MVD = −3 maps to code_num 6.

6.4.13.2 Context-Based Adaptive Variable Length Coding (CAVLC)

This is the method used to encode residual, zig-zag ordered 4 × 4 (and 2 × 2) blocks of transform coefficients. CAVLC [6] is designed to take advantage of several characteristics of quantised 4 × 4 blocks:

1. After prediction, transformation and quantisation, blocks are typically sparse (containing mostly zeros). CAVLC uses run-level coding to represent strings of zeros compactly.
2. The highest nonzero coefficients after the zig-zag scan are often sequences of ±1 and CAVLC signals the number of high-frequency ±1 coefficients ('Trailing Ones') in a compact way.
3. The number of nonzero coefficients in neighbouring blocks is correlated. The number of coefficients is encoded using a look-up table and the choice of look-up table depends on the number of nonzero coefficients in neighbouring blocks.
4. The level (magnitude) of nonzero coefficients tends to be larger at the start of the reordered array (near the DC coefficient) and smaller towards the higher frequencies. CAVLC takes advantage of this by adapting the choice of VLC look-up table for the level parameter depending on recently-coded level magnitudes.

CAVLC encoding of a block of transform coefficients proceeds as follows:

coeff_token	encodes the number of non-zero coefficients (TotalCoeff) and TrailingOnes (one per block)
trailing_ones_sign_flag	sign of TrailingOne value (one per trailing one)
level_prefix	first part of code for non-zero coefficient (one per coefficient, excluding trailing ones)
level_suffix	second part of code for non-zero coefficient (not always present)
total_zeros	encodes the total number of zeros occurring after the first non-zero coefficient (in zig-zag order) (one per block)
run_before	encodes number of zeros preceding each non-zero coefficient *in reverse zig-zag order*

1. Encode the number of coefficients and trailing ones (coeff_token)

The first VLC, coeff_token, encodes both the total number of nonzero coefficients (TotalCoeffs) and the number of trailing ±1 values (TrailingOnes). TotalCoeffs can be anything from 0 (no coefficients in the 4 × 4 block)[5] to 16 (16 nonzero coefficients) and TrailingOnes can be anything from 0 to 3. If there are more than three trailing ±1s, only the last three are treated as 'special cases' and any others are coded as normal coefficients.

There are four choices of look-up table to use for encoding coeff_token for a 4 × 4 block, three variable-length code tables and a fixed-length code table. The choice of table depends on the number of nonzero coefficients in the left-hand and upper previously coded blocks (n_A and n_B respectively). A parameter nC is calculated as follows. If upper and left blocks nB and nA

---

[5] Note: coded_block_pattern (described earlier) indicates which *8 × 8* blocks in the macroblock contain nonzero coefficients but, within a coded 8 × 8 block, there may be *4 × 4* sub-blocks that do not contain any coefficients, hence TotalCoeff may be 0 in any 4 × 4 sub-block. In fact, this value of TotalCoeff occurs most often and is assigned the shortest VLC.

Table 6.12 Choice of look-up table for
coeff_token

N	Table for coeff_token
0, 1	Table 1
2, 3	Table 2
4, 5, 6, 7	Table 3
8 or above	Table 4

are both available (i.e. in the same coded slice), nC = round((nA + nB)/2). If only the upper
is available, nC = nB; if only the left block is available, nC = nA; if neither is available,
nC = 0.

The parameter nC selects the look-up table (Table 6.12) so that the choice of VLC
adapts to the number of coded coefficients in neighbouring blocks (*context adaptive*). Table 1
is biased towards small numbers of coefficients such that low values of TotalCoeffs are
assigned particularly short codes and high values of TotalCoeff particularly long codes.
Table 2 is biased towards medium numbers of coefficients (TotalCoeff values around 2–4
are assigned relatively short codes), Table 3 is biased towards higher numbers of coefficients and Table 4 assigns a fixed six-bit code to every pair of TotalCoeff and TrailingOnes
values.

2. Encode the sign of each TrailingOne
For each TrailingOne (trailing ±1) signalled by coeff_token, the sign is encoded with a single
bit (0 = +, 1 = −) *in reverse order*, starting with the highest-frequency TrailingOne.

3. Encode the levels of the remaining nonzero coefficients.
The *level* (sign and magnitude) of each remaining nonzero coefficient in the block is encoded *in
reverse order*, starting with the highest frequency and working back towards the DC coefficient.
The code for each level is made up of a prefix (level_prefix) and a suffix (level_suffix). The
length of the suffix (suffixLength) may be between 0 and 6 bits and suffixLength is adapted
depending on the magnitude of each successive coded level ('context adaptive'). A small
value of suffixLength is appropriate for levels with low magnitudes and a larger value of
suffixLength is appropriate for levels with high magnitudes. The choice of suffixLength is
adapted as follows:

1. Initialise suffixLength to 0 (unless there are more than 10 nonzero coefficients and less
   than three trailing ones, in which case initialise to 1).
2. Encode the highest-frequency nonzero coefficient.
3. If the magnitude of this coefficient is larger than a predefined threshold, increment suffixLength. (If this is the first level to be encoded and suffixLength was initialised to 0, set
   suffixLength to 2).

In this way, the choice of suffix (and hence the complete VLC) is matched to the magnitude of
the recently-encoded coefficients. The thresholds are listed in Table 6.13; the first threshold is

Table 6.13 Thresholds for determining whether to increment suffixLength

Current suffixLength	Threshold to increment suffixLength
0	0
1	3
2	6
3	12
4	24
5	48
6	N/A (highest suffixLength)

zero which means that suffixLength is always incremented after the first coefficient level has been encoded.

4. Encode the total number of zeros before the last coefficient

The sum of all zeros preceding the highest nonzero coefficient in the reordered array is coded with a VLC, total zeros. The reason for sending a separate VLC to indicate total zeros is that many blocks contain a number of nonzero coefficients at the start of the array and (as will be seen later) this approach means that zero-runs at the start of the array need not be encoded.

5. Encode each run of zeros.

The number of zeros preceding each nonzero coefficient (run_before) is encoded *in reverse order*. A run_before parameter is encoded for each nonzero coefficient, starting with the highest frequency, with two exceptions:

1. If there are no more zeros left to encode (i.e. $\sum[\text{run\_before}] = \text{total\_zeros}$), it is not necessary to encode any more run_before values.
2. It is not necessary to encode run_before for the final (lowest frequency) nonzero coefficient.

The VLC for each run of zeros is chosen depending on (a) the number of zeros that have not yet been encoded (ZerosLeft) and (b) run_before. For example, if there are only two zeros left to encode, run_before can only take three values (0, 1 or 2) and so the VLC need not be more than two bits long. If there are six zeros still to encode then run_before can take seven values (0 to 6) and the VLC table needs to be correspondingly larger.

Example 1

4 × 4 block:

0	3	−1	0
0	−1	1	0
1	0	0	0
0	0	0	0

Reordered block:
0,3,0,1,−1, −1,0,1,0...

TotalCoeffs = 5 (indexed from highest frequency, 4, to lowest frequency, 0)
total_zeros = 3
TrailingOnes = 3 (in fact there are four trailing ones but only three can be encoded as a 'special case')

Encoding

Element	Value	Code
coeff_token	TotalCoeffs = 5, TrailingOnes= 3 (use Table 1)	0000100
TrailingOne sign (4)	+	0
TrailingOne sign (3)	−	1
TrailingOne sign (2)	−	1
Level (1)	+1 (use suffixLength = 0)	1 (prefix)
Level (0)	+3 (use suffixLength = 1)	001 (prefix) 0 (suffix)
total zeros	3	111
run_before(4)	ZerosLeft = 3; run_before = 1	10
run_before(3)	ZerosLeft = 2; run_before = 0	1
run_before(2)	ZerosLeft = 2; run_before = 0	1
run_before(1)	ZerosLeft = 2; run_before = 1	01
run_before(0)	ZerosLeft = 1; run_before = 1	No code required; last coefficient.

The transmitted bitstream for this block is 0000100011100101111101101.

Decoding
The output array is 'built up' from the decoded values as shown below. Values added to the output array at each stage are underlined.

Code	Element	Value	Output array
0000100	coeff_token	TotalCoeffs = 5, TrailingOnes = 3	Empty
0	TrailingOne sign	+	$\underline{1}$
1	TrailingOne sign	−	$\underline{-1}$, 1
1	TrailingOne sign	−	$\underline{-1}$, −1, 1
1	Level	+1 (suffixLength = 0; increment suffixLength after decoding)	$\underline{1}$, −1, −1, 1
0010	Level	+3 (suffixLength = 1)	$\underline{3}$, 1, −1, −1, 0, 1
111	total_zeros	3	3, 1, −1, −1, 1
10	run_before	1	3, 1, −1, −1, $\underline{0}$, 1
1	run_before	0	3, 1, −1, −1, 0, 1
1	run_before	0	3, 1, −1, −1, 0, 1
01	run_before	1	3, $\underline{0}$, 1, −1, −1, 0, 1

The decoder has already inserted two zeros, TotalZeros is equal to 3 and so another 1 zero is inserted before the lowest coefficient, making the final output array:

$$\underline{0}, 3, 0, 1, -1, -1, 0, 1$$

Example 2

4×4 block:

−2	4	0	−1
3	0	0	0
−3	0	0	0
0	0	0	0

Reordered block:
−2, 4, 3, −3, 0, 0, −1, ...

TotalCoeffs = 5 (indexed from highest frequency, 4, to lowest frequency, 0)
total_zeros = 2
TrailingOne = 1

Encoding:

Element	Value	Code
coeff_token	TotalCoeffs = 5, TrailingOnes = 1 (use Table 1)	0000000110
TrailingOne sign (4)	−	1
Level (3)	Sent as −2 (*see note* 1) (suffixLength = 0; increment suffixLength)	0001 (prefix)
Level (2)	3 (suffixLength = 1)	001 (prefix) 0 (suffix)
Level (1)	4 (suffixLength = 1; increment suffixLength	0001 (prefix) 0 (suffix)
Level (0)	−2 (suffixLength = 2)	1 (prefix) 11 (suffix)
total zeros	2	0011
run_before(4)	ZerosLeft= 2; run_before= 2	00
run_before(3..0)	0	No code required

The transmitted bitstream for this block is 00000001101000100100001011001100.

Note 1: Level (3), with a value of −3, is encoded as a special case. If there are less than 3 TrailingOnes, then the first *non*-trailing one level cannot have a value of ±1 (otherwise it would have been encoded as a TrailingOne). To save bits, this level is incremented if negative (decremented if positive) so that ±2 maps to ±1, ±3 maps to ±2, and so on. In this way, shorter VLCs are used.

Note 2: After encoding level (3), the level_VLC table is incremented because the magnitude of this level is greater than the first threshold (which is 0). After encoding level (1), with a magnitude of 4, the table number is incremented again because level (1) is greater than the second threshold (which is 3). Note that the final level (−2) uses a different VLC from the first encoded level (also −2).

Decoding:

Code	Element	Value	Output array
0000000110	coeff_token	TotalCoeffs = 5, T1s= 1	Empty
1	TrailingOne sign	—	−1
0001	Level	−2 decoded as −3	−3, −1
0010	Level	+3	+3, −3, −1
00010	Level	+4	+4, 3, −3, −1
111	Level	−2	−2, 4, 3, −3, −1
0011	total_zeros	2	−2, 4, 3, −3, −1
00	run_before	2	−2, 4, 3, −3, 0, 0, −1

All zeros have now been decoded and so the output array is:
−2, 4, 3, −3, 0, 0, −1
(This example illustrates how bits are saved by encoding TotalZeros: only a single zero run
(run_before) needs to be coded even though there are five nonzero coefficients.)

Example 3

4 × 4 block:

0	0	1	0
0	0	0	0
1	0	0	0
−0	0	0	0

Reordered block:
0,0,0,1,0,1,0,0,0,−1

TotalCoeffs = 3 (indexed from highest frequency [2] to lowest frequency [0])
total_zeros = 7
TrailingOnes = 3

Encoding:

Element	Value	Code
coeff_token	TotalCoeffs = 3, TrailingOnes= 3 use Table 1)	00011
TrailingOne sign (2)	—	1
TrailingOne sign (1)	+	0
TrailingOne sign (0)	+	0
total_zeros	7	011
run_before(2)	ZerosLeft= 7; run_before= 3	100
run_before(1)	ZerosLeft= 4; run_before= 1	10
run_before(0)	ZerosLeft= 3; run_before= 3	No code required; last coefficient.

The transmitted bitstream for this block is 0001110001110010.

Decoding:

Code	Element	Value	Output array
00011	coeff_token	TotalCoeffs= 3, TrailingOnes= 3	Empty
1	TrailineOne sign	−	−1
0	TrailineOne sign	+	1, −1
0	TrailineOne sign	+	1, 1, −1
011	total_zeros	7	1, 1, −1
100	run_before	3	1, 1, 0, 0, 0, −1
10	run_before	1	1, 0, 1, 0, 0, 0, −1

The decoder has inserted four zeros. total_zeros is equal to 7 and so another three zeros are inserted before the lowest coefficient:

$$0, 0, 0, 1, 0, 1, 0, 0, 0, -1$$

6.5 THE MAIN PROFILE

Suitable application for the Main Profile include (but are not limited to) broadcast media applications such as digital television and stored digital video. The Main Profile is almost a superset of the Baseline Profile, except that multiple slice groups, ASO and redundant slices (all included in the Baseline Profile) are not supported. The additional tools provided by Main Profile are B slices (bi-predicted slices for greater coding efficiency), weighted prediction (providing increased flexibility in creating a motion-compensated prediction block), support for interlaced video (coding of fields as well as frames) and CABAC (an alternative entropy coding method based on Arithmetic Coding).

6.5.1 B slices

Each macroblock partition in an inter coded macroblock in a B slice may be predicted from one or two reference pictures, before or after the current picture in temporal order. Depending on the reference pictures stored in the encoder and decoder (see the next section), this gives many options for choosing the prediction references for macroblock partitions in a B macroblock type. Figure 6.40 shows three examples: (a) one past and one future reference (similar to B-picture prediction in earlier MPEG video standards), (b) two past references and (c) two future references.

6.5.1.1 Reference pictures

B slices use two lists of previously-coded reference pictures, list 0 and list 1, containing short term and long term pictures (see Section 6.4.2). These two lists can each contain past and/or

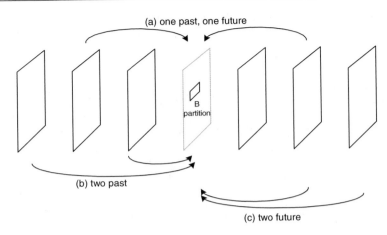

Figure 6.40 Partition prediction examples in a B macroblock type: (a) past/future, (b) past, (c) future

future coded pictures (pictures before or after the current picture in display order). The long term pictures in each list behaves in a similar way to the description in Section 6.4.2. The short term pictures may be past and/or future coded pictures and the default index order of these pictures is as follows:

List 0: The closest past picture (based on picture order count) is assigned index 0, followed by any other past pictures (increasing in picture order count), followed by any future pictures (in increasing picture order count from the current picture).

List 1: The closest future picture is assigned index 0, followed by any other future picture (in increasing picture order count), followed by any past picture (in increasing picture order count).

Example

An H.264 decoder stores six short term reference pictures with picture order counts: 123, 125, 126, 128, 129, 130. The current picture is 127. All six short term reference pictures are marked as used for reference in list 0 and list 1. The pictures are indexed in the list 0 and list 1 short term buffers as follows (Table 6.14).

Table 6.14 Short term buffer indices (B slice prediction) (current picture order count is 127

Index	List 0	List 1
0	126	128
1	125	129
2	123	130
3	128	126
4	129	125
5	130	123

Table 6.15 Prediction options in B slice macroblocks

Partition	Options
16 × 16	Direct, list 0, list1 or bi-predictive
16 × 8 or 8 × 16	List 0, list 1 or bi-predictive (chosen separately for each partition)
8 × 8	Direct, list 0, list 1 or bi-predictive (chosen separately for each partition).

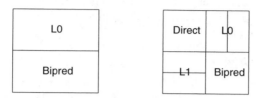

Figure 6.41 Examples of prediction modes in B slice macroblocks

The selected buffer index is sent as an Exp-Golomb codeword (see Section 6.4.13.1) and so the most efficient choice of reference index (with the smallest codeword) is index 0 (i.e. the previous coded picture in list 0 and the next coded picture in list 1).

6.5.1.2 Prediction Options

Macroblocks partitions in a B slice may be predicted in one of several ways, direct mode (see Section 6.5.1.4), motion-compensated prediction from a list 0 reference picture, motion-compensated prediction from a list 1 reference picture, or motion-compensated bi-predictive prediction from list 0 and list 1 reference pictures (see Section 6.5.1.3). Different prediction modes may be chosen for each partition (Table 6.15); if the 8 × 8 partition size is used, the chosen mode for each 8 × 8 partition is applied to all sub-partitions within that partition. Figure 6.41 shows two examples of valid prediction mode combinations. On the left, two 16 × 8 partitions use List 0 and Bi-predictive prediction respectively and on the right, four 8 × 8 partitions use Direct, List 0, List 1 and Bi-predictive prediction.

6.5.1.3 Bi-prediction

In Bi-predictive mode, a reference block (of the same size as the current partition or sub-macroblock partition) is created from the list 0 and list 1 reference pictures. Two motion-compensated reference areas are obtained from a list 0 and a list 1 picture respectively (and hence two motion vectors are required) and each sample of the prediction block is calculated as an average of the list 0 and list 1 prediction samples. Except when using Weighted Prediction (see Section 6.5.2), the following equation is used:

$$\text{pred}(i,j) = (\text{pred0}(i,j) + \text{pred1}(i,j) + 1) >> 1$$

Pred0(i, j) and pred1(i, j) are prediction samples derived from the list 0 and list 1 reference frames and pred(i, j) is a bi-predictive sample. After calculating each prediction sample, the motion-compensated residual is formed by subtracting pred(i, j) from each sample of the current macroblock as usual.

Example

A macroblock is predicted in B_Bi_16 × 16 mode (i.e. bi-prediction of the complete macroblock). Figure 6.42 and Figure 6.43 show motion-compensated reference areas from list 0 and list 1 references pictures respectively and Figure 6.44 shows the bi-prediction formed from these two reference areas.

The list 0 and list 1 vectors in a bi-predictive macroblock or block are each predicted from neighbouring motion vectors that have the same temporal direction. For example a vector for the current macroblock pointing to a past frame is predicted from other neighbouring vectors that also point to past frames.

6.5.1.4 Direct Prediction

No motion vector is transmitted for a B slice macroblock or macroblock partition encoded in Direct mode. Instead, the decoder calculates list 0 and list 1 vectors based on previously-coded vectors and uses these to carry out bi-predictive motion compensation of the decoded residual samples. A skipped macroblock in a B slice is reconstructed at the decoder using Direct prediction.

A flag in the slice header indicates whether a spatial or temporal method will be used to calculate the vectors for direct mode macroblocks or partitions.

In *spatial direct* mode, list 0 and list 1 predicted vectors are calculated as follows. Predicted list 0 and list 1 vectors are calculated using the process described in section 6.4.5.3. If the co-located MB or partition in the first list 1 reference picture has a motion vector that is less than $\pm\frac{1}{2}$ luma samples in magnitude (and in some other cases), one or both of the predicted vectors are set to zero; otherwise the predicted list 0 and list 1 vectors are used to carry out bi-predictive motion compensation. In *temporal direct* mode, the decoder carries out the following steps:

1. Find the list 0 reference picture for the co-located MB or partition in the list 1 picture. This list 0 reference becomes the list 0 reference of the current MB or partition.
2. Find the list 0 vector, MV, for the co-located MB or partition in the list 1 picture.
3. Scale vector MV based on the picture order count 'distance' between the current and list 1 pictures: this is the new list 1 vector MV1.
4. Scale vector MV based on the picture order count distance between the current and list 0 pictures: this is the new list 0 vector MV0.

These modes are modified when, for example, the prediction reference macroblocks or partitions are not available or are intra coded.

Example:

The list 1 reference for the current macroblock occurs two pictures after the current frame (Figure 6.45). The co-located MB in the list 1 reference has a vector MV(+2.5, +5) pointing to a list 0 reference picture that occurs three pictures before the current picture. The decoder calculates MV1(−1, −2) and MV0(+1.5, +3) pointing to the list 1 and list 0 pictures respectively. These vectors are derived from MV and have magnitudes proportional to the picture order count distance to the list 0 and list 1 reference frames.

Figure 6.42 Reference area (list 0 picture)

Figure 6.43 Reference area (list 1 picture)

Figure 6.44 Prediction (non-weighted)

6.5.2 Weighted Prediction

Weighted prediction is a method of modifying (scaling) the samples of motion-compensated prediction data in a P or B slice macroblock. There are three types of weighted prediction in H.264:

1. P slice macroblock, 'explicit' weighted prediction;
2. B slice macroblock, 'explicit' weighted prediction;
3. B slice macroblock, 'implicit' weighted prediction.

Each prediction sample $pred0(i, j)$ or $pred1(i, j)$ is scaled by a weighting factor w_0 or w_1 prior to motion-compensated prediction. In the 'explicit' types, the weighting factor(s) are

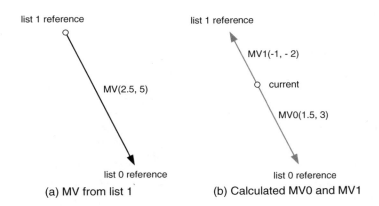

Figure 6.45 Temporal direct motion vector example

determined by the encoder and transmitted in the slice header. If 'implicit' prediction is used, w_0 and w_1 are calculated based on the relative temporal positions of the list 0 and list 1 reference pictures. A larger weighting factor is applied if the reference picture is temporally close to the current picture and a smaller factor is applied if the reference picture is temporally further away from the current picture.

One application of weighted prediction is to allow explicit or implicit control of the relative contributions of reference picture to the motion-compensated prediction process. For example, weighted prediction may be effective in coding of 'fade' transitions (where one scene fades into another).

6.5.3 Interlaced Video

Efficient coding of interlaced video requires tools that are optimised for compression of field macroblocks. If field coding is supported, the type of picture (frame or field) is signalled in the header of each slice. In macroblock-adaptive frame/field (**MB-AFF**) coding mode, the choice of field or frame coding may be specified at the macroblock level. In this mode, the current slice is processed in units of 16 luminance samples wide and 32 luminance samples high, each of which is coded as a 'macroblock pair' (Figure 6.46). The encoder can choose to encode each MB pair as (a) two frame macroblocks or (b) two field macroblocks and may select the optimum coding mode for each region of the picture.

Coding a slice or MB pair in field mode requires modifications to a number of the encoding and decoding steps described in Section 6.4. For example, each coded field is treated as a separate reference picture for the purposes of P and B slice prediction, the prediction of coding modes in intra MBs and motion vectors in inter MBs require to be modified depending on whether adjacent MBs are coded in frame or field mode and the reordering scan shown in Figure 6.47 replaces the zig-zag scan of Figure 6.39.

6.5.4 Context-based Adaptive Binary Arithmetic Coding (CABAC)

When the picture parameter set flag entropy_coding_mode is set to 1, an arithmetic coding system is used to encode and decode H.264 syntax elements. Context-based Adaptive Binary

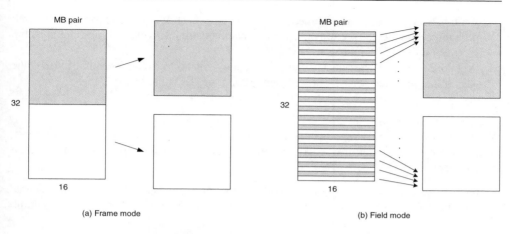

Figure 6.46 Macroblock-adaptive frame/field coding

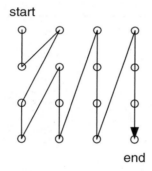

Figure 6.47 Reordering scan for 4 × 4 luma blocks (field mode)

Arithmetic Coding (CABAC) [7], achieves good compression performance through (a) selecting probability models for each syntax element according to the element's context, (b) adapting probability estimates based on local statistics and (c) using arithmetic coding rather than variable-length coding. Coding a data symbol involves the following stages:

1. Binarisation: CABAC uses Binary Arithmetic Coding which means that only binary decisions (1 or 0) are encoded. A non-binary-valued symbol (e.g. a transform coefficient or motion vector, any symbol with more than 2 possible values) is 'binarised' or converted into a binary code prior to arithmetic coding. This process is similar to the process of converting a data symbol into a variable length code (Section 6.4.13) but the binary code is further encoded (by the arithmetic coder) prior to transmission.

Stages 2, 3 and 4 are repeated for each bit (or 'bin') of the binarised symbol:

2. Context model selection. A 'context model' is a probability model for one or more bins of the binarised symbol and is chosen from a selection of available models depending on the statistics of recently-coded data symbols. The context model stores the probability of each bin being '1' or '0'.

3. Arithmetic encoding: An arithmetic coder encodes each bin according to the selected probability model (see section 3.5.3). Note that there are just two sub-ranges for each bin (corresponding to '0' and '1').

4. Probability update: The selected context model is updated based on the actual coded value (e.g. if the bin value was '1', the frequency count of '1's is increased).

The Coding Process

We will illustrate the coding process for one example, mvd_x (motion vector difference in the x-direction, coded for each partition or sub-macroblock partition in an inter macroblock).

1. Binarise the value $mvd_x \cdots mvd_x$ is mapped to the following table of uniquely-decodeable codewords for $|mvd_x| < 9$ (larger values of mvd_x are binarised using an Exp-Golomb codeword).

$|mvd_x|$	Binarisation (s=sign)
0	0
1	10s
2	110s
3	1110s
4	11110s
5	111110s
6	1111110s
7	11111110s
8	111111110s

The first bit of the binarised codeword is bin 1, the second bit is bin 2 and so on.

2. Choose a *context model* for each bin. One of three models is selected for bin 1 (Table 6.16), based on the L1 norm of two previously-coded mvd_x values, e_k:

$$e_k = |mvd_{xA}| + |mvd_{xB}| \quad \text{where A and B are the blocks immediately to the left and above the current block.}$$

If e_k is small, then there is a high probability that the current MVD will have a small magnitude and, conversely, if e_k is large then it is more likely that the current MVD will have a large magnitude. A probability table (context model) is selected accordingly. The remaining bins are coded using one of four further context models (Table 6.17).

Table 6.16 context models for bin 1

e_k	Context model for bin 1
$0 \le e_k < 3$	Model 0
$3 \le e_k < 33$	Model 1
$33 \le e_k$	Model 2

Table 6.17 Context models

Bin	Context model
1	0, 1 or 2 depending on e_k
2	3
3	4
4	5
5 and higher	6
6	6

3. Encode each bin. The selected context model supplies two probability estimates, the probability that the bin contains '1' and the probability that the bin contains '0', that determine the two sub-ranges used by the arithmetic coder to encode the bin.
4. Update the context models. For example, if context model 2 is selected for bin 1 and the value of bin 1 is '0', the frequency count of '0's is incremented so that the next time this model is selected, the probability of an '0' will be slightly higher. When the total number of occurrences of a model exceeds a threshold value, the frequency counts for '0' and '1' will be scaled down, which in effect gives higher priority to recent observations.

The Context Models

Context models and binarisation schemes for each syntax element are defined in the standard. There are nearly 400 separate context models for the various syntax elements. At the beginning of each coded slice, the context models are initialised depending on the initial value of the Quantisation Parameter QP (since this has a significant effect on the probability of occurrence of the various data symbols). In addition, for coded P, SP and B slices, the encoder may choose one of 3 sets of context model initialisation parameters at the beginning of each slice, to allow adaptation to different types of video content [8].

The Arithmetic Coding Engine

The arithmetic decoder is described in some detail in the Standard and has three distinct properties:

1. Probability estimation is performed by a transition process between 64 separate probability states for 'Least Probable Symbol' (LPS, the least probable of the two binary decisions '0' or '1').
2. The range R representing the current state of the arithmetic coder (see Chapter 3) is quantised to a small range of pre-set values before calculating the new range at each step, making it possible to calculate the new range using a look-up table (i.e. multiplication-free).
3. A simplified encoding and decoding process (in which the context modelling part is bypassed) is defined for data symbols with a near-uniform probability distribution.

The definition of the decoding process is designed to facilitate low-complexity implementations of arithmetic encoding and decoding. Overall, CABAC provides improved coding efficiency compared with VLC (see Chapter 7 for performance examples).

6.6 THE EXTENDED PROFILE

The Extended Profile (known as the X Profile in earlier versions of the draft H.264 standard) may be particularly useful for applications such as video streaming. It includes all of the features of the Baseline Profile (i.e. it is a superset of the Baseline Profile, unlike Main Profile), together with B-slices (Section 6.5.1), Weighted Prediction (Section 6.5.2) and additional features to support efficient streaming over networks such as the Internet. SP and SI slices facilitate switching between different coded streams and 'VCR-like' functionality and Data Partitioned slices can provide improved performance in error-prone transmission environments.

6.6.1 SP and SI slices

SP and SI slices are specially-coded slices that enable (among other things) efficient switching between video streams and efficient random access for video decoders [10]. A common requirement in a streaming application is for a video decoder to switch between one of several encoded streams. For example, the same video material is coded at multiple bitrates for transmission across the Internet and a decoder attempts to decode the highest-bitrate stream it can receive but may require switching automatically to a lower-bitrate stream if the data throughput drops.

Example

A decoder is decoding Stream A and wants to switch to decoding Stream B (Figure 6.48). For simplicity, assume that each frame is encoded as a single slice and predicted from one reference (the previous decoded frame). After decoding P-slices A_0 and A_1, the decoder wants to switch to Stream B and decode B_2, B_3 and so on. If all the slices in Stream B are coded as P-slices, then the decoder will not have the correct decoded reference frame(s) required to reconstruct B_2 (since B_2 is predicted from the decoded picture B_1 which does not exist in stream A). One solution is to code frame B_2 as an I-slice. Because it is coded without prediction from any other frame, it can be decoded independently of preceding frames in stream B and the decoder can therefore switch between stream A and stream B as shown in Figure 6.49. Switching can be accommodated by inserting an I-slice at regular intervals in the coded sequence to create 'switching points'. However, an I-slice is likely to contain much more coded data than a P-slice and the result is an undesirable peak in the coded bitrate at each switching point.

SP-slices are designed to support switching between similar coded sequences (for example, the same source sequence encoded at various bitrates) without the increased bitrate penalty of I-slices (Figure 6.49). At the switching point (frame 2 in each sequence), there are three SP-slices, each coded using motion compensated prediction (making them more efficient than I-slices). SP-slice A_2 can be decoded using reference picture A_1 and SP-slice B_2 can be decoded using reference picture B_1. The key to the switching process is SP-slice AB_2 (known as a *switching SP-slice*), created in such a way that it can be decoded using motion-compensated reference picture A_1, to produce decoded frame B_2 (i.e. the decoder output frame B_2 is identical whether decoding B_1 followed by B_2 or A_1 followed by AB_2). An extra SP-slice is required at each switching point (and in fact another SP-slice, BA_2, would be required to switch in the other direction) but this is likely to be more efficient than encoding frames A_2

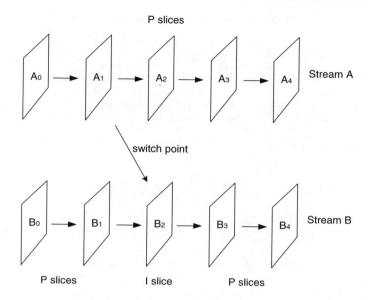

Figure 6.48 Switching streams using I-slices

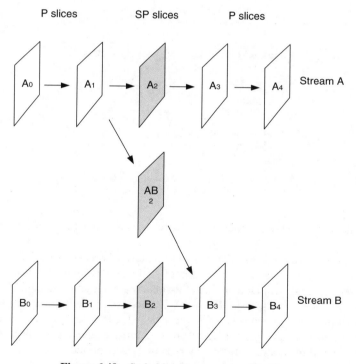

Figure 6.49 Switching streams using SP-slices

Table 6.18 Switching from stream A to stream B using SP-slices

Input to decoder	MC reference	Output of decoder
P-slice A_0	[earlier frame]	Decoded frame A_0
P-slice A_1	Decoded frame A_0	Decoded frame A_1
SP-slice AB_2	Decoded frame A_1	Decoded frame B_2
P-slice B_3	Decoded frame B_2	Decoded frame B_3
....	....	....

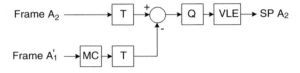

Figure 6.50 Encoding SP-slice A_2 (simplified)

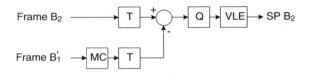

Figure 6.51 Encoding SP-slice B_2 (simplified)

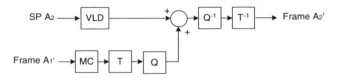

Figure 6.52 Decoding SP-slice A_2 (simplified)

and B_2 as I-slices. Table 6.18 lists the steps involved when a decoder switches from stream A to stream B.

Figure 6.50 shows a simplified diagram of the encoding process for SP-slice A_2, produced by subtracting a motion-compensated version of A_1' (decoded frame A_1) from frame A_2 and then coding the residual. Unlike a 'normal' P-slice, the subtraction occurs in the transform domain (after the block transform). SP-slice B_2 is encoded in the same way (Figure 6.51). A decoder that has previously decoded frame A_1 can decode SP-slice A_2 as shown in Figure 6.52. Note that these diagrams are simplified; in practice further quantisation and rescaling steps are required to avoid mismatch between encoder and decoder and a more detailed treatment of the process can be found in [11].

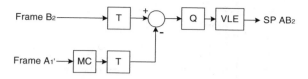

Figure 6.53   Encoding SP-slice AB$_2$ (simplified)

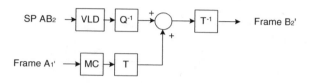

Figure 6.54   Decoding SP-slice AB$_2$ (simplified)

SP-slice AB$_2$ is encoded as shown in Figure 6.53 (simplified). Frame B$_2$ (the frame we are switching to) is transformed and a motion-compensated prediction is formed from A$_1'$ (the decoded frame from which we are switching). The 'MC' block in this diagram attempts to find the best match for each MB of frame B$_2$ *using decoded picture A$_1$ as a reference.* The motion-compensated prediction is transformed, then subtracted from the transformed B$_2$ (i.e. in the case of a switching SP slice, subtraction takes place in the transform domain). The residual (after subtraction) is quantized, encoded and transmitted.

A decoder that has previously decoded A$_1'$ can decode SP-slice AB$_2$ to produce B$_2'$ (Figure 6.54). A$_1'$ is motion compensated (using the motion vector data encoded as part of AB$_2$), transformed and added to the decoded and scaled (inverse quantized) residual, then the result is inverse transformed to produce B$_2'$.

If streams A and B are versions of the same original sequence coded at different bitrates, the motion-compensated prediction of B$_2$ from A$_1'$ (SP-slice AB$_2$) should be quite efficient. Results show that using SP-slices to switch between different versions of the same sequence is significantly more efficient than inserting I-slices at switching points. Another application of SP-slices is to provide random access and 'VCR-like' functionalities. For example, an SP-slice and a switching SP-slice are placed at the position of frame 10 (Figure 6.55). A decoder can fast-forward from A$_0$ directly to A$_{10}$ by (a) decoding A$_0$, then (b) decoding switching SP-slice A$_{0-10}$ to produce A$_{10}$ by prediction from A$_0$.

A further type of switching slice, the SI-slice, is supported by the Extended Profile. This is used in a similar way to a switching SP-slice, except that the prediction is formed using the 4 × 4 Intra Prediction modes (see Section 6.4.6.1) from previously-decoded samples of the reconstructed frame. This slice mode may be used (for example) to switch from one sequence to a completely different sequence (in which case it will not be efficient to use motion compensated prediction because there is no correlation between the two sequences).

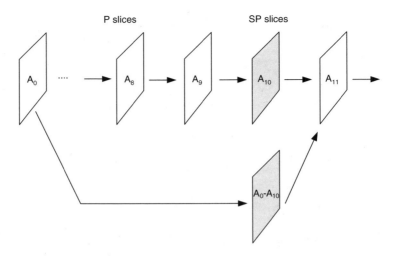

Figure 6.55 Fast-forward using SP-slices

Sequence parameter set	SEI	Picture parameter set	I slice	Picture delimiter	P slice	P slice	·····

Figure 6.56 Example sequence of RBSP elements

6.6.2 Data Partitioned Slices

The coded data that makes up a slice is placed in three separate Data Partitions (A, B and C), each containing a subset of the coded slice. Partition A contains the slice header and header data for each macroblock in the slice, Partition B contains coded residual data for Intra and SI slice macroblocks and Partition C contains coded residual data for inter coded macroblocks (forward and bi-directional). Each Partition can be placed in a separate NAL unit and may therefore be transported separately.

If Partition A data is lost, it is likely to be difficult or impossible to reconstruct the slice, hence Partition A is highly sensitive to transmission errors. Partitions B and C can (with careful choice of coding parameters) be made to be independently decodeable and so a decoder may (for example) decode A and B only, or A and C only, lending flexibility in an error-prone environment.

6.7 TRANSPORT OF H.264

A coded H.264 video sequence consists of a series of NAL units, each containing an RBSP (Table 6.19). Coded slices (including Data Partitioned slices and IDR slices) and the End of Sequence RBSP are defined as VCL NAL units whilst all other elements are just NAL units.

An example of a typical sequence of RBSP units is shown in Figure 6.56. Each of these units is transmitted in a separate NAL unit. The header of the NAL unit (one byte) signals the type of RBSP unit and the RBSP data makes up the rest of the NAL unit.

Table 6.19

RBSP type	Description
Parameter Set	'Global' parameters for a sequence such as picture dimensions, video format, macroblock allocation map (see Section 6.4.3).
Supplemental Enhancement Information	Side messages that are not essential for correct decoding of the video sequence.
Picture Delimiter	Boundary between video pictures (optional). If not present, the decoder infers the boundary based on the frame number contained within each slice header.
Coded slice	Header and data for a slice; this RBSP unit contains actual coded video data.
Data Partition A, B or C	Three units containing Data Partitioned slice layer data (useful for error resilient decoding). Partition A contains header data for all MBs in the slice, Partition B contains intra coded data and partition C contains inter coded data.
End of sequence	Indicates that the next picture (in decoding order) is an IDR picture (see Section 6.4.2). (Not essential for correct decoding of the sequence).
End of stream	Indicates that there are no further pictures in the bitstream. (Not essential for correct decoding of the sequence).
Filler data	Contains 'dummy' data (may be used to increase the number of bytes in the sequence). (Not essential for correct decoding of the sequence).

Parameter sets

H.264 introduces the concept of *parameter sets*, each containing information that can be applied to a large number of coded pictures. A *sequence parameter set* contains parameters to be applied to a complete video sequence (a set of consecutive coded pictures). Parameters in the sequence parameter set include an identifier (seq_parameter_set_id), limits on frame numbers and picture order count, the number of reference frames that may be used in decoding (including short and long term reference frames), the decoded picture width and height and the choice of progressive or interlaced (frame or frame / field) coding. A *picture parameter set* contains parameters which are applied to one or more decoded pictures within a sequence. Each picture parameter set includes (among other parameters) an identifier (pic_parameter_set_id), a selected seq_parameter_set_id, a flag to select VLC or CABAC entropy coding, the number of slice groups in use (and a definition of the type of slice group map), the number of reference pictures in list 0 and list 1 that may be used for prediction, initial quantizer parameters and a flag indicating whether the default deblocking filter parameters are to be modified.

Typically, one or more sequence parameter set(s) and picture parameter set(s) are sent to the decoder prior to decoding of slice headers and slice data. A coded slice header refers to a pic_parameter_set_id and this 'activates' that particular picture parameter set. The 'activated' picture parameter set then remains active until a different picture parameter set is activated by being referred to in another slice header. In a similar way, a picture parameter set refers to a seq_parameter_set_id which 'activates' that sequence parameter set. The activated set remains in force (i.e. its parameters are applied to all consecutive coded pictures) until a different sequence parameter set is activated.

The parameter set mechanism enables an encoder to signal important, infrequently-changing sequence and picture parameters separately from the coded slices themselves. The parameter sets may be sent well ahead of the slices that refer to them, or by another transport mechanism (e.g. over a reliable transmission channel or even 'hard wired' in a decoder implementation). Each coded slice may 'call up' the relevant picture and sequence parameters using a single VLC (pic_parameter_set_id) in the slice header.

Transmission and Storage of NAL units

The method of transmitting NAL units is not specified in the standard but some distinction is made between transmission over packet-based transport mechanisms (e.g. packet networks) and transmission in a continuous data stream (e.g. circuit-switched channels). In a packet-based network, each NAL unit may be carried in a separate packet and should be organised into the correct sequence prior to decoding. In a circuit-switched transport environment, a start code prefix (a uniquely-identifiable delimiter code) is placed before each NAL unit to make a *byte stream* prior to transmission. This enables a decoder to search the stream to find a start code prefix identifying the start of a NAL unit.

In a typical application, coded video is required to be transmitted or stored together with associated audio track(s) and side information. It is possible to use a range of transport mechanisms to achieve this, such as the Real Time Protocol and User Datagram Protocol (RTP/UDP). An Amendment to MPEG-2 Systems specifies a mechanism for transporting H.264 video (see Chapter 7) and ITU-T Recommendation H.241 defines procedures for using H.264 in conjunction with H.32× multimedia terminals. Many applications require storage of multiplexed video, audio and side information (e.g. streaming media playback, DVD playback). A forthcoming Amendment to MPEG-4 Systems (Part 1) specifies how H.264 coded data and associated media streams can be stored in the ISO Media File Format (see Chapter 7).

6.8 CONCLUSIONS

H.264 provides mechanisms for coding video that are optimised for compression efficiency and aim to meet the needs of practical multimedia communication applications. The range of available coding tools is more restricted than MPEG-4 Visual (due to the narrower focus of H.264) but there are still many possible choices of coding parameters and strategies. The success of a practical implementation of H.264 (or MPEG-4 Visual) depends on careful design of the CODEC and effective choices of coding parameters. The next chapter examines design issues for each of the main functional blocks of a video CODEC and compares the performance of MPEG-4 Visual and H.264.

6.9 REFERENCES

1. ISO/IEC 14496-10 and ITU-T Rec. H.264, Advanced Video Coding, 2003.
2. T. Wiegand, G. Sullivan, G. Bjontegaard and A. Luthra, Overview of the H.264 / AVC Video Coding Standard, IEEE Transactions on Circuits and Systems for Video Technology, to be published in 2003.
3. A. Hallapuro, M. Karczewicz and H. Malvar, Low Complexity Transform and Quantization – Part I: Basic Implementation, JVT document JVT-B038, Geneva, February 2002.
4. H.264 Reference Software Version JM6.1d, http://bs.hhi.de/~suehring/tml/, March 2003.

5. S. W. Golomb, Run-length encoding, *IEEE* Trans. on Inf. Theory, **IT-12**, pp. 399–401, 1966.

6. G. Bjøntegaard and K. Lillevold, Context-adaptive VLC coding of coefficients, JVT document JVT-C028, Fairfax, May 2002.

7. D. Marpe, G. Blättermann and T. Wiegand, Adaptive codes for H.26L, ITU-T SG16/6 document VCEG-L13, Eibsee, Germany, January 2001.

8. H. Schwarz, D. Marpe and T. Wiegand, CABAC and slices, JVT document JVT-D020, Klagenfurt, Austria, July 2002

9. D. Marpe, H. Schwarz and T. Wiegand, Context-Based Adaptive Binary Arithmetic Coding in the H.264 / AVC Video Compression Standard, IEEE Transactions on Circuits and Systems for Video Technology, to be published in 2003.

10. M. Karczewicz and R. Kurceren, A proposal for SP-frames, ITU-T SG16/6 document VCEG-L27, Eibsee, Germany, January 2001.

11. M. Karczewicz and R. Kurceren, The SP and SI Frames Design for H.264/AVC, IEEE Transactions on Circuits and Systems for Video Technology, to be published in 2003.

7

Design and Performance

7.1 INTRODUCTION

The MPEG-4 Visual and H.264 standards include a range of coding tools and processes and there is significant scope for differences in the way standards-compliant encoders and decoders are developed. Achieving good performance in a practical implementation requires careful design and careful choice of coding parameters.

In this chapter we give an overview of practical issues related to the design of software or hardware implementations of the coding standards. The design of each of the main functional blocks of a CODEC (such as motion estimation, transform and entropy coding) can have a significant impact on computational efficiency and compression performance. We discuss the interfaces to a video encoder and decoder and the value of video pre-processing to reduce input noise and post-processing to minimise coding artefacts.

Comparing the performance of video coding algorithms is a difficult task, not least because decoded video quality is dependent on the input video material and is inherently subjective. We compare the subjective and objective (PSNR) coding performance of MPEG-4 Visual and H.264 reference model encoders using selected test video sequences. Compression performance often comes at a computational cost and we discuss the computational performance requirements of the two standards.

The compressed video data produced by an encoder is typically stored or transmitted across a network. In many practical applications, it is necessary to control the bitrate of the encoded data stream in order to match the available bitrate of a delivery mechanism. We discuss practical bitrate control and network transport issues.

7.2 FUNCTIONAL DESIGN

Figures 3.51 and 3.52 show typical structures for a motion-compensated transform based video encoder and decoder. A practical MPEG-4 Visual or H.264 CODEC is required to implement some or all of the functions shown in these figures (even if the CODEC structure is different

---

H.264 and MPEG-4 Video Compression: Video Coding for Next-generation Multimedia.
Iain E. G. Richardson.   © 2003 John Wiley & Sons, Ltd. ISBN: 0-470-84837-5

from that shown). Conforming to the MPEG-4/H.264 standards, whilst maintaining good compression and computational performance, requires careful design of the CODEC functional blocks. The goal of a functional block design is to achieve good rate/distortion performance (see Section 7.4.3) whilst keeping computational overheads to an acceptable level.

Functions such as motion estimation, transforms and entropy coding can be highly computationally intensive. Many practical platforms for video compression are power-limited or computation-limited and so it is important to design the functional blocks with these limitations in mind. In this section we discuss practical approaches and tradeoffs in the design of the main functional blocks of a video CODEC.

7.2.1 Segmentation

The object-based functionalities of MPEG-4 (Core, Main and related profiles) require a video scene to be *segmented* into objects. Segmentation methods usually fall into three categories:

1. Manual segmentation: this requires a human operator to identify manually the borders of each object in each source video frame, a very time-consuming process that is obviously only suitable for 'offline' video content (video data captured in advance of coding and transmission). This approach may be appropriate, for example, for segmentation of an important visual object that may be viewed by many users and/or re-used many times in different composed video sequences.
2. Semi-automatic segmentation: a human operator identifies objects and perhaps object boundaries in one frame; a segmentation algorithm refines the object boundaries (if necessary) and tracks the video objects through successive frames of the sequence.
3. Fully-automatic segmentation: an algorithm attempts to carry out a complete segmentation of a visual scene without any user input, based on (for example) spatial characteristics such as edges and temporal characteristics such as object motion between frames.

Semi-automatic segmentation [1,2] has the potential to give better results than fully-automatic segmentation but still requires user input. Many algorithms have been proposed for automatic segmentation [3,4]. In general, better segmentation performance can be achieved at the expense of greater computational complexity. Some of the more sophisticated segmentation algorithms require significantly more computation than the video encoding process itself. Reasonably accurate segmentation performance can be achieved by spatio-temporal approaches (e.g. [3]) in which a coarse approximate segmentation is formed based on spatial information and is then refined as objects move. Excellent segmentation results can be obtained in controlled environments (for example, if a TV presenter stands in front of a blue background) but the results for practical scenarios are less robust.

The output of a segmentation process is a sequence of mask frames for each VO, each frame containing a binary mask for one VOP (e.g. Figure 5.30) that determines the processing of MBs and blocks and is coded as a BAB in each boundary MB position.

7.2.2 Motion Estimation

Motion estimation is the process of selecting an offset to a suitable reference area in a previously coded frame (see Chapter 3). Motion estimation is carried out in a video encoder (not in a

32×32 block in current frame

Figure 7.1 Current block (white border)

decoder) and has a significant effect on CODEC performance. A good choice of prediction reference minimises the energy in the motion-compensated residual which in turn maximises compression performance. However, finding the 'best' offset can be a very computationally-intensive procedure.

The offset between the current region or block and the reference area (motion vector) may be constrained by the semantics of the coding standard. Typically, the reference area is constrained to lie within a rectangle centred upon the position of the current block or region. Figure 7.1 shows a 32×32-sample block (outlined in white) that is to be motion-compensated. Figure 7.2 shows the same block position in the previous frame (outlined in white) and a larger square extending ± 7 samples around the block position in each direction. The motion vector may 'point' to any reference area within the larger square (the search area). The goal of a practical motion estimation algorithm is to find a vector that minimises the residual energy after motion compensation, whilst keeping the computational complexity within acceptable limits. The choice of algorithm depends on the platform (e.g. software or hardware) and on whether motion estimation is block-based or region-based.

7.2.2.1 Block Based Motion Estimation

Energy Measures

Motion compensation aims to minimise the energy of the residual transform coefficients after quantisation. The energy in a transformed block depends on the energy in the residual block (prior to the transform). Motion estimation therefore aims to find a 'match' to the current block or region that minimises the energy in the motion compensated residual (the difference between the current block and the reference area). This usually involves evaluating the residual energy at a number of different offsets. The choice of measure for 'energy' affects computational complexity and the accuracy of the motion estimation process. Equation 7.1, equation 7.2 and equation 7.3 describe three energy measures, MSE, MAE and SAE. The motion

Previous (reference) frame

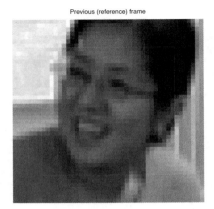

Figure 7.2   Search region in previous (reference) frame

compensation block size is $N \times N$ samples; C_{ij} and R_{ij} are current and reference area samples respectively.

1. Mean Squared Error:
$$MSE = \frac{1}{N^2} \sum_{i=0}^{N-1} \sum_{j=0}^{N-1} (C_{ij} - R_{ij})^2 \qquad (7.1)$$

2. Mean Absolute Error:
$$MAE = \frac{1}{N^2} \sum_{i=0}^{N-1} \sum_{j=0}^{N-1} |C_{ij} - R_{ij}| \qquad (7.2)$$

3. Sum of Absolute Errors:
$$SAE = \sum_{i=0}^{N-1} \sum_{j=0}^{N-1} |C_{ij} - R_{ij}| \qquad (7.3)$$

Example

Evaluating MSE for every possible offset in the search region of Figure 7.2 gives a 'map' of MSE (Figure 7.3). This graph has a minimum at $(+2, 0)$ which means that the best match is obtained by selecting a 32×32 sample reference region at an offset of 2 to the right of the block position in the current frame. MAE and SAE (sometimes referred to as SAD, Sum of Absolute Differences) are easier to calculate than MSE; their 'maps' are shown in Figure 7.4 and Figure 7.5. Whilst the gradient of the map is different from the MSE case, both these measures have a minimum at location $(+2, 0)$.

SAE is probably the most widely-used measure of residual energy for reasons of computational simplicity. The H.264 reference model software [5] uses SA(T)D, the sum of absolute differences of the *transformed* residual data, as its prediction energy measure (for both Intra and Inter prediction). Transforming the residual at each search location increases computation but improves the accuracy of the energy measure. A simple multiply-free transform is used and so the extra computational cost is not excessive.

The results of the above example indicate that the best choice of motion vector is $(+2,0)$. The minimum of the MSE or SAE map indicates the offset that produces a minimal residual energy and this is likely to produce the smallest energy of quantised transform

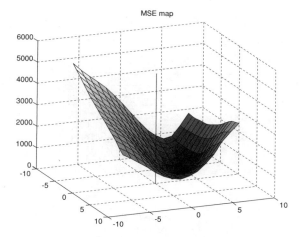

Figure 7.3   MSE map

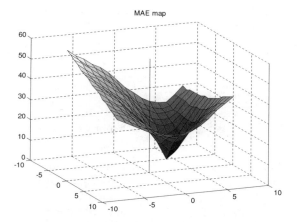

Figure 7.4   MAE map

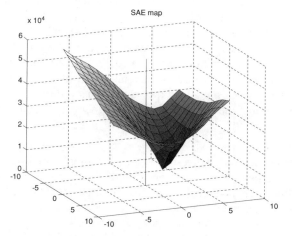

Figure 7.5   SAE map

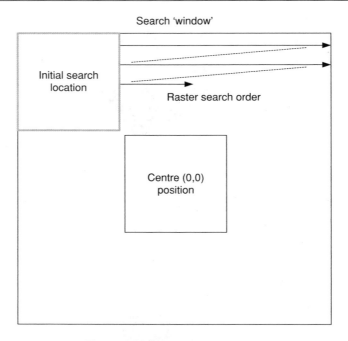

Figure 7.6 Full search (raster scan)

coefficients. The motion vector itself must be transmitted to the decoder, however, and as larger vectors are coded using more bits than small-magnitude vectors (see Chapter 3) it may be useful to 'bias' the choice of vector towards (0,0). This can be achieved simply by subtracting a constant from the MSE or SAE at position (0,0). A more sophisticated approach is to treat the choice of vector as a constrained optimisation problem [6]. The H.264 reference model encoder [5] adds a cost parameter for each coded element (MVD, prediction mode, etc) before choosing the smallest total cost of motion prediction.

It may not always be necessary to calculate SAE (or MAE or MSE) completely at each off-set location. A popular shortcut is to terminate the calculation early once the previous minimum SAE has been exceeded. For example, after calculating each inner sum of equation (7.3) ($\sum_{j=0}^{N-1} |C_{ij} - R_{ij}|$), the encoder compares the total SAE with the previous minimum. If the total so far exceeds the previous minimum, the calculation is terminated (since there is no point in finishing the calculation if the outcome is already higher than the previous minimum SAE).

Full Search

Full Search motion estimation involves evaluating equation 7.3 (SAE) at each point in the search window ($\pm S$ samples about position (0,0), the position of the current macroblock). Full search estimation is guaranteed to find the minimum SAE (or MAE or MSE) in the search window but it is computationally intensive since the energy measure (e.g. equation (7.3)) must be calculated at every one of $(2S + 1)^2$ locations.

Figure 7.6 shows an example of a Full Search strategy. The first search location is at the top-left of the window (position [$-S$, $-S$]) and the search proceeds in raster order

Search 'window'

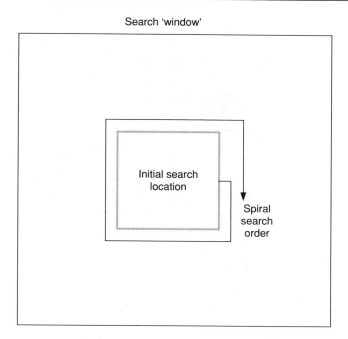

Figure 7.7 Full search (spiral scan)

until all positions have been evaluated. In a typical video sequence, most motion vectors are concentrated around (0,0) and so it is likely that a minimum will be found in this region. The computation of the full search algorithm can be simplified by starting the search at (0,0) and proceeding to test points in a spiral pattern around this location (Figure 7.7). If early termination is used (see above), the SAE calculation is increasingly likely to be terminated early (thereby saving computation) as the search pattern widens outwards.

'Fast' Search Algorithms

Even with the use of early termination, Full Search motion estimation is too computationally intensive for many practical applications. In computation- or power-limited applications, so-called 'fast search' algorithms are preferable. These algorithms operate by calculating the energy measure (e.g. SAE) at a subset of locations within the search window.

The popular Three Step Search (TSS, sometimes described as N-Step Search) is illustrated in Figure 7.8. SAE is calculated at position (0,0) (the centre of the Figure) and at eight locations $\pm 2^{N-1}$ (for a search window of $\pm(2^N - 1)$ samples). In the figure, S is 7 and the first nine search locations are numbered '1'. The search location that gives the smallest SAE is chosen as the new search centre and a further eight locations are searched, this time at half the previous distance from the search centre (numbered '2' in the figure). Once again, the 'best' location is chosen as the new search origin and the algorithm is repeated until the search distance cannot be subdivided further. The TSS is considerably simpler than Full Search ($8N + 1$ searches compared with $(2^{N+1} - 1)^2$ searches for Full Search) but the TSS (and other fast

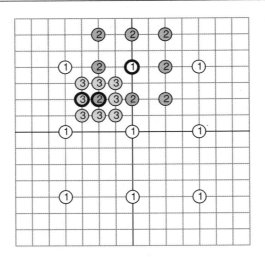

Figure 7.8   Three Step Search

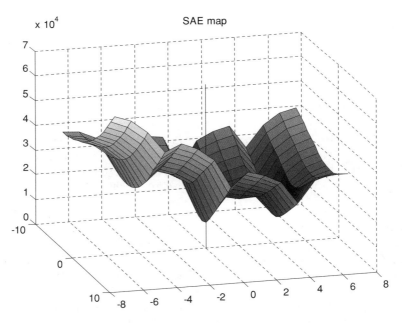

Figure 7.9   SAE map showing several local minima

search algorithms) do not usually perform as well as Full Search. The SAE map shown in Figure 7.5 has a single minimum point and the TSS is likely to find this minimum correctly, but the SAE map for a block containing complex detail and/or different moving components may have several local minima (e.g. see Figure 7.9). Whilst the Full Search will always identify the global minimum, a fast search algorithm may become 'trapped' in a local minimum, giving a suboptimal result.

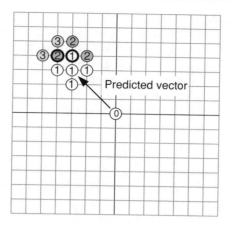

Figure 7.10 Nearest Neighbours Search

Many fast search algorithms have been proposed, such as Logarithmic Search, Hierarchical Search, Cross Search and One at a Time Search [7–9]. In each case, the performance of the algorithm can be evaluated by comparison with Full Search. Suitable comparison criteria are compression performance (how effective is the algorithm at minimising the motion-compensated residual?) and computational performance (how much computation is saved compared with Full Search?). Other criteria may be helpful; for example, some 'fast' algorithms such as Hierarchical Search are better-suited to hardware implementation than others.

Nearest Neighbours Search [10] is a fast motion estimation algorithm that has low computational complexity but closely approaches the performance of Full Search within the framework of MPEG-4 Simple Profile. In MPEG-4 Visual, each block or macroblock motion vector is differentially encoded. A predicted vector is calculated (based on previously-coded vectors from neighbouring blocks) and the difference (MVD) between the current vector and the predicted vector is transmitted. NNS exploits this property by giving preference to vectors that are close to the predicted vector (and hence minimise MVD). First, SAE is evaluated at location (0,0). Then, the search origin is set to the predicted vector location and surrounding points in a diamond shape are evaluated (labelled '1' in Figure 7.10). The next step depends on which of the points have the lowest SAE. If the (0,0) point or the centre of the diamond have the lowest SAE, then the search terminates. If a point on the edge of the diamond has the lowest SAE (the highlighted point in this example), that becomes the centre of a new diamond-shaped search pattern and the search continues. In the figure, the search terminates after the points marked '3' are searched. The inherent bias towards the predicted vector gives excellent compression performance (close to the performance achieved by full search) with low computational complexity.

Sub-pixel Motion Estimation

Chapter 3 demonstrated that better motion compensation can be achieved by allowing the offset into the reference frame (the motion vector) to take fractional values rather than just integer values. For example, the woman's head will not necessarily move by an integer number of pixels from the previous frame (Figure 7.2) to the current frame (Figure 7.1). Increased fractional accuracy (half-pixel vectors in MPEG-4 Simple Profile, quarter-pixel vectors in Advanced Simple profile and H.264) can provide a better match and reduce the energy in the motion-compensated residual. This gain is offset against the need to transmit fractional motion vectors (which increases the number of bits required to represent motion vectors) and the increased complexity of sub-pixel motion estimation and compensation.

Sub-pixel motion estimation requires the encoder to interpolate between integer sample positions in the reference frame as discussed in Chapter 3. Interpolation is computationally intensive, especially so for quarter-pixel interpolation because a high-order interpolation filter is required for good compression performance (see Chapter 6). Calculating sub-pixel samples for the entire search window is not usually necessary. Instead, it is sufficient to find the best integer-pixel match (using Full Search or one of the fast search algorithms discussed above) and then to search interpolated positions adjacent to this position. In the case of quarter-pixel motion estimation, first the best integer match is found; then the best half-pixel position match in the immediate neighbourhood is calculated; finally the best quarter-pixel match around this half-pixel position is found.

7.2.2.2 Object Based Motion Estimation

Chapter 5 described the process of motion compensated prediction and reconstruction (MC/MR) of boundary MBs in an MPEG-4 Core Profile VOP. During MC/MR, transparent pixels in boundary and transparent MBs are padded prior to forming a motion compensated prediction. In order to find the optimum prediction for each MB, motion estimation should be carried out using the padded reference frame. Object-based motion estimation consists of the following steps.

1. Pad transparent pixel positions in the reference VOP as described in Chapter 5.
2. Carry out block-based motion estimation to find the best match for the current MB in the padded reference VOP. If the current MB is a boundary MB, the energy measure should only be calculated for opaque pixel positions in the current MB.

Motion estimation for arbitrary-shaped VOs is more complex than for rectangular frames (or slices/VOs). In [11] the computation and compression performance of a number of popular motion estimation algorithms are compared for the rectangular and object-based cases. Methods of padding boundary MBs using graphics co-processor functions are described in [12] and a hardware architecture for Motion Estimation, Motion Compensation and CAE shape coding is presented in [13].

7.2.3 DCT/IDCT

The Discrete Cosine Transform is widely used in image and video compression algorithms in order to decorrelate image or residual data prior to quantisation and compression (see Chapter 3). The basic FDCT and IDCT equations (equations (3.4) and (3.5)), if implemented directly, require a large number of multiplications and additions. It is possible to exploit the structure of the transform matrix **A** in order to significantly reduce computational complexity and this is one of the reasons for the popularity of the DCT.

7.2.3.1 8 × 8 DCT

Direct evaluation of equation (3.4) for an 8 × 8 FDCT (where $N = 8$) requires $64 \times 64 = 4096$ multiplications and accumulations. From the matrix form (equation (3.1)) it is clear that the 2D transform can be evaluated in two stages (i.e. calculate $\mathbf{AX}$ and then multiply by matrix $\mathbf{A}^T$, or vice versa). The 1D FDCT is given by equation (7.4), where f_i are the N input samples and F_x are the N output coefficients. Rearranging the 2D FDCT equation (equation (3.4)) shows that the 2D FDCT can be constructed from two 1D transforms (equation (7.5)). The 2D FDCT may be calculated by evaluating a 1D FDCT of each column of the input matrix (the inner transform), then evaluating a 1D FDCT of each row of the result of the first set of transforms (the outer transform). The 2D IDCT can be manipulated in a similar way (equation (7.6)). Each eight-point 1D transform takes 64 multiply/accumulate operations, giving a total of $64 \times 8 \times 2 = 1024$ multiply/accumulate operations for an 8 × 8 FDCT or IDCT.

$$F_x = C_x \sum_{i=0}^{N-1} f_i \cos \frac{(2i+1)x\pi}{2N} \tag{7.4}$$

$$Y_{xy} = C_x \sum_{i=0}^{N-1} \left[C_y \sum_{j=0}^{N-1} X_{ij} \cos \frac{(2j+1)y\pi}{2N} \right] \cos \frac{(2i+1)x\pi}{2N} \tag{7.5}$$

$$X_{ij} = \sum_{x=0}^{N-1} C_x \left[\sum_{y=0}^{N-1} C_y Y_{xy} \cos \frac{(2j+1)y\pi}{2N} \right] \cos \frac{(2i+1)x\pi}{2N} \tag{7.6}$$

At first glance, calculating an eight-point 1-D FDCT (equation (7.4)) requires the evaluation of eight different cosine factors ($\cos \frac{(2i+1)x\pi}{2N}$ with eight values of i) for each of eight coefficient indices ($x = 0\ldots7$). However, the symmetries of the cosine function make it possible to combine many of these calculations into a reduced number of steps. For example, consider the calculation of F_2 (from equation (7.4)):

$$F_2 = \frac{1}{2} \left[f_0 \cos\left(\frac{\pi}{8}\right) + f_1 \cos\left(\frac{3\pi}{8}\right) + f_2 \cos\left(\frac{5\pi}{8}\right) + f_3 \cos\left(\frac{7\pi}{8}\right) + f_4 \cos\left(\frac{9\pi}{8}\right) \right.$$
$$\left. + f_5 \cos\left(\frac{11\pi}{8}\right) + f_6 \cos\left(\frac{13\pi}{8}\right) + f_7 \cos\left(\frac{15\pi}{8}\right) \right] \tag{7.7}$$

Evaluating equation (7.7) would seem to require eight multiplications and seven additions (plus a scaling by a half). However, by making use of the symmetrical properties of the cosine function this can be simplified to:

$$F_2 = \frac{1}{2} \left[(f_0 - f_4 + f_7 - f_3).\cos\left(\frac{\pi}{8}\right) + (f_1 - f_2 - f_5 + f_6).\cos\left(\frac{3\pi}{8}\right) \right] \tag{7.8}$$

In a similar way, F_6 may be simplified to:

$$F_6 = \frac{1}{2} \left[(f_0 - f_4 + f_7 - f_3).\cos\left(\frac{3\pi}{8}\right) + (f_1 - f_2 - f_5 + f_6).\cos\left(\frac{\pi}{8}\right) \right] \tag{7.9}$$

The additions and subtractions are common to both coefficients and need only be carried out once so that F_2 and F_6 can be calculated using a total of eight additions and four multiplications (plus a final scaling by a half). Extending this approach to the complete 8 × 8 FDCT leads to a number of alternative 'fast' implementations such as the popular algorithm due to

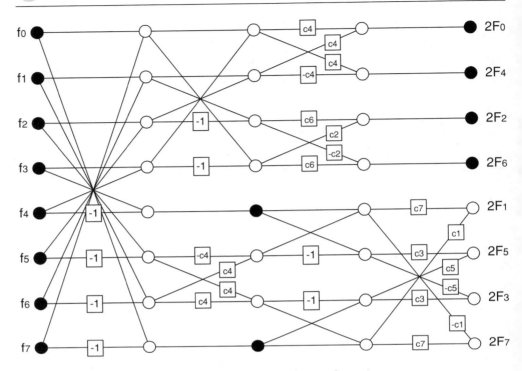

Figure 7.11   FDCT flowgraph

Chen, Smith and Fralick [14]. The data flow through this 1D algorithm can be represented as a 'flowgraph' (Figure 7.11). In this figure, a circle indicates addition of two inputs, a square indicates multiplication by a constant and cX indicates the constant $\cos(X\pi/16)$. This algorithm requires only 26 additions or subtractions and 20 multiplications (in comparison with the 64 multiplications and 64 additions required to evaluate equation (7.4)).

Figure 7.11 is just one possible simplification of the 1D DCT algorithm. Many flowgraph-type algorithms have been developed over the years, optimised for a range of implementation requirements (e.g. minimal multiplications, minimal subtractions, etc.) Further computational gains can be obtained by direct optimisation of the 2D DCT (usually at the expense of increased implementation complexity).

Flowgraph algorithms are very popular for software CODECs where (in many cases) the best performance is achieved by minimising the number of computationally-expensive multiply operations. For a hardware implementation, regular data flow may be more important than the number of operations and so a different approach may be required. Popular hardware architectures for the FDCT / IDCT include those based on parallel multiplier arrays and distributed arithmetic [15–18].

7.2.3.2 H.264 4 × 4 Transform

The integer IDCT approximations specified in the H.264 standard have been designed to be suitable for fast, efficient software and hardware implementation. The original proposal for the

Figure 7.12 8 × 8 block in boundary MB

forward and inverse transforms [19] describes alternative implementations using (i) a series of shifts and additions ('shift and add'), (ii) a flowgraph algorithm and (iii) matrix multiplications. Some platforms (for example DSPs) are better-suited to 'multiply-accumulate' calculations than to 'shift and add' operations and so the matrix implementation (described in C code in [20]) may be more appropriate for these platforms.

7.2.3.3 Object Boundaries

In a Core or Main Profile MPEG-4 CODEC, residual coefficients in a boundary MB are coded using the 8 × 8 DCT. Figure 7.12 shows one block from a boundary MB (with the transparent pixels set to 0 and displayed here as black). The entire block (including the transparent pixels) is transformed with an 8 × 8 DCT and quantised and the reconstructed block after rescaling and inverse DCT is shown in Figure 7.13. Note that some of the formerly transparent pixels are now nonzero due to quantisation distortion (e.g. the pixel marked with a white 'cross'). The decoder discards the transparent pixels (according to the BAB transparency map) and retains the opaque pixels.

Using an 8 × 8 DCT and IDCT for an irregular-shaped region of opaque pixels is not ideal because the transparent pixels contribute to the energy in the DCT coefficients and so more data is coded than is absolutely necessary. Because the transparent pixel positions are discarded by the decoder, the encoder may place any data at all in these positions prior to the DCT. Various strategies have been proposed for filling (padding) the transparent positions prior to applying the 8 × 8 DCT, for example, by padding with values selected to minimise the energy in the DCT coefficients [21, 22], but choosing the optimal padding values is a computationally expensive process. A simple alternative is to pad the transparent positions in an inter-coded MB with zeros (since the motion-compensated residual is usually close to zero anyway) and to pad the transparent positions in an inter-coded MB with the value 2^{N-1}, where N is the number of bits per pixel (since this is mid-way between the minimum and maximum pixel value). The Shape-Adaptive DCT (see Chapter 5) provides a more efficient solution for transforming irregular-shaped blocks but is computationally intensive and is only available in the Advanced Coding Efficiency Profile of MPEG-4 Visual.

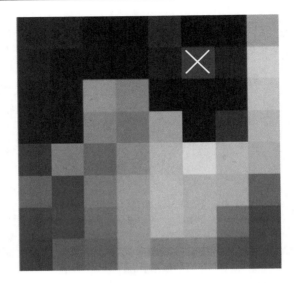

Figure 7.13   8 × 8 block after FDCT, quant, rescale, IDCT

7.2.4 Wavelet Transform

The DWT was chosen for MPEG-4 still texture coding because it can out-perform block-based transforms for still image coding (although the Intra prediction and transform in H.264 performs well for still images). A number of algorithms have been proposed for the efficient coding and decoding of the DWT [23–25]. One issue related to software and hardware implementations of the DWT is that it requires substantially more memory than block transforms, since the transform operates on a complete image or a large section of an image (rather than a relatively small block of samples).

7.2.5 Quantise/Rescale

Scalar quantisation and rescaling (Chapter 3) can be implemented by division and/or multiplication by constant parameters (controlled by a quantisation parameter or quantiser step size). In general, multiplication is an expensive computation and some gains may be achieved by integrating the quantisation and rescaling multiplications with the forward and inverse transforms respectively. In H.264, the specification of the quantiser is combined with that of the transform in order to facilitate this combination (see Chapter 6).

7.2.6 Entropy Coding

7.2.6.1 Variable-Length Encoding

In Chapter 3 we introduced the concept of entropy coding using variable-length codes (VLCs). In MPEG-4 Visual and H.264, the VLC required to encode each data symbol is defined by the standard. During encoding each data symbol is replaced by the appropriate VLC, determined by (a) the context (e.g. whether the data symbol is a header value, transform coefficient,

Table 7.1 Variable-length encoding example

Value, V	Length, L	Value	Size	Value	Size	Output
–	–	–	0	–	0	–
101	3	101	3	101	3	–
11100	5	**11100101**	8	–	0	**11100101**
100	3	100	3	100	3	–
101	3	101100	6	101100	6	–
101	3	**101101100**	9	1	1	**01101100**
11100	5	111001	6	111001	6	–
1101	4	**1101111001**	10	11	2	**01111001**
. . . etc.						

Column groups: Input VLC (Value, V — Length, L), R (before output) (Value — Size), R (after output) (Value — Size), Output

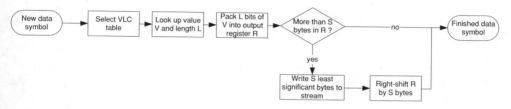

Figure 7.14 Variable length encoding flowchart

motion vector component, etc.) and (b) the value of the data symbol. Chapter 3 presented some examples of pre-defined VLC tables from MPEG-4 Visual.

VLCs (by definition) contain variable numbers of bits but in many practical transport situations it is necessary to map a series of VLCs produced by the encoder to a stream of bytes or words. A mechanism for carrying this out is shown in Figure 7.14. An output register, R, collects encoded VLCs until enough data are present to write out one or more bytes to the stream. When a new data symbol is encoded, the value V of the VLC is concatenated with the previous contents of R (with the new VLC occupying the most significant bits). A count of the number of bits held in R is incremented by L (the length of the new VLC in bits). If R contains more than S bytes (where S is the number of bytes to be written to the stream at a time), the S least significant bytes of R are written to the stream and the contents of R are right-shifted by S bytes.

Example

A series of VLCs (from Table 3.12, Chapter 3) are encoded using the above method. $S = 1$, i.e. 1 byte is written to the stream at a time. Table 7.1 shows the variable-length encoding process at each stage with each output byte highlighted in bold type.

Figure 7.15 shows a basic architecture for carrying out the VLE process. A new data symbol and context indication (table selection) are passed to a look-up unit that returns the value V and length L of the codeword. A packer unit concatenates sequences of VLCs and outputs S bytes at a time (in a similar way to the above example).

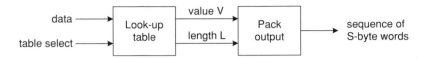

Figure 7.15 Variable length encoding architecture

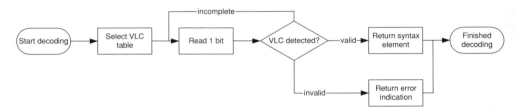

Figure 7.16 Flowchart for decoding one VLC

Issues to consider when designing a variable length encoder include computational efficiency and look-up table size. In software, VLE can be processor-intensive because of the large number of bit-level operations required to pack and shift the codes. Look-up table design can be problematic because of the large size and irregular structure of VLC tables. For example, the MPEG-4 Visual TCOEF table (see Chapter 3) is indexed by the three parameters Run (number of preceding zero coefficients), Level (nonzero coefficient level) and Last (final nonzero coefficient in a block). There are only 102 valid VLCs but over 16 000 valid combinations of Run, Level and Last, each corresponding to a VLC of up to 13 bits or a 20-bit 'Escape' code, and so this table may require a significant amount of storage. In the H.264 Variable Length Coding scheme, many symbols are represented by 'universal' Exp-Golomb codes that can be calculated from the data symbol value (avoiding the need for large VLC look-up tables) (see Chapter 6).

7.2.6.2 Variable-length Decoding

Decoding VLCs involves 'scanning' or parsing a received bitstream for valid codewords, extracting these codewords and decoding the appropriate syntax elements. As with the encoding process, it is necessary for the decoder to know the current context in order to select the correct codeword table. Figure 7.16 illustrates a simple method of decoding one VLC. The decoder reads successive bits of the input bitstream until a valid VLC is detected (the usual case) or an invalid VLC is detected (i.e. a code that is not valid within the current context). For example, a code starting with nine or more zeros is not a valid VLC if the decoder is expecting an MPEG-4 Transform Coefficient. The decoder returns the appropriate syntax element if a valid VLC is found, or an error indication if an invalid VLC is detected.

VLC decoding can be computationally intensive, memory intensive or both. One method of implementing the decoder is as a Finite State Machine. The decoder starts at an initial state and moves through successive states based on the value of each bit. Eventually, the decoder reaches a state that corresponds to (a) a complete, valid VLC or (b) an invalid VLC. The

Table 7.2   Variable length decoding example: MPEG-4 Visual TCOEF

State	Input	Next state	VLC	Output (last, run, level)
0	0	1	0..	–
	1	2	1..	–
1	0	..later state	00..	
	1	..later state	01..	
2	0	0	**10s**	(0,0,s1)
	1	3	11..	
3	0	0	**110s**	(0,1,s1)
	1	4	111..	–
4	0	0	**1110s**	(0,2,s1)
	1	0	**1111s**	(0,0,s2)
etc.	...	...	...	...

decoded syntax element (or error indication) is returned and the decoder restarts from the initial state. Table 7.2 shows the first part of the decoding sequence for the MPEG-4 Visual TCOEF (transform coefficient) context, starting with state 0. If the input bit is 0, the next state is state 1 and if the input is 1, the next state is 2. From state 2, if the input is 0, the decoder has 'found' a valid VLC, 10. In this context it is necessary to decode 1 more bit at the end of each VLC (the sign bit, s, indicating whether the level is positive or negative), after which it outputs the relevant syntax element (0, 1 or $+/-1$ in this case) and returns to state 0. Note that when a syntax element containing 'last $= 1$' is decoded, we have reached the end of the block of coefficients and it is necessary to reset or change the context.

In this example, the decoder can process one input bit at each stage (e.g. one bit per clock cycle in a hardware implementation). This may be too slow for some applications in which case a more sophisticated architecture that can examine multiple bits (or entire VLCs) in one operation may be required. Examples of architectures for variable-length coding and decoding include [26–29].

7.2.6.3 Arithmetic Coding

An arithmetic encoder (see Chapter 3) encodes each syntax element through successive refinement of a fractional number. Arithmetic coding has the potential for greater compression efficiency than any variable-length coding algorithm (due to its ability to represent fractional probability distributions accurately). In practice, it is usually necessary to represent the fractional numbers produced by an arithmetic encoder using fixed-point values within a limited dynamic range. Some implementation issues for the context-based arithmetic coder adopted for H.264 Main Profile are discussed in Chapter 6 and a detailed overview of the CABAC scheme is given in [30].

7.3 INPUT AND OUTPUT

7.3.1 Interfacing

Figure 7.17 shows a system in which video frames are encoded, transmitted or stored and decoded. At the input to the encoder (A) and the output of the decoder (D), data are in the

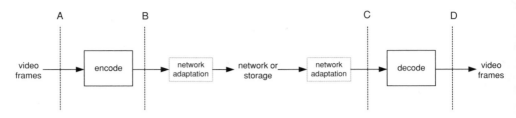

Figure 7.17   Video CODEC interfaces

format of uncompressed video frames, each represented by a set of samples, typically in the YCbCr colour space using one of the sampling structures described in Chapter 2 (4:4:4, 4:2:2 or 4:2:0). There are a number of different methods of combining the three components of each frame, including interleaved (samples of Y, Cb and Cr are interleaved together in raster scan order), concatenated (the complete Y component for a frame is followed by the Cb and then Cr components) and using separate buffers or memory areas to store each of the three components. The choice of method may depend on the application. For example, using separate buffers for the Y, Cb and Cr components may be suitable for a software CODEC; a hardware CODEC with limited memory and/or a requirement for low delay may use an interleaved format.

At the output of the encoder (B) and the input to the decoder (C) the data consist of a sequence of bits representing the video sequence in coded form. The H.264 and MPEG-4 Visual standards use fixed length codes, variable-length codes and/or arithmetic coding to represent the syntax elements of the compressed sequence. The coded bitstream consists of continuous sequences of bits, interspersed with fixed-length 'marker' codes. Methods of mapping this bitstream to a transport or storage mechanism ('delivery mechanism') include the following.

Bit-oriented: If the delivery mechanism is capable of dealing with an arbitrary number of bits, the bitstream may be transmitted directly (optionally multiplexed with associated data such as coded audio and 'side' information).

Byte-oriented: Many delivery mechanisms (e.g. file storage or network packets) require data to be mapped to an integral number of bytes or words. It may be necessary to pad the coded data at the end of a unit (e.g. slice, picture, VOP or sequence) to make an integral number of bytes or words.

Packet-oriented: Both MPEG-4 Visual and H.264 support the concept of placing a complete coded unit in a network packet. A video packet or NAL unit packet contains coded data that corresponds to a discrete coded unit such as a slice (a complete frame or VOP or a portion of a frame or VOP) (see Section 6.7).

7.3.2 Pre-processing

The compression efficiency of a video CODEC can be significantly improved by pre-processing video frames prior to encoding. Problems with the source material and/or video capture system may degrade the coding performance of a video encoder. *Camera noise* (introduced by the camera and/or the digitisation process) is illustrated in Figure 7.18. The top

Figure 7.18   Image showing camera noise (lower half)

half of this image is relatively noise-free and this is typical of the type of image captured by a high-quality digital camera. Images captured from low-quality sources are more likely to contain noise (shown in the lower half of this figure). Camera noise may appear in higher spatial frequencies and change from frame to frame. An encoder will 'see' this noise as a high-frequency component that is present in the motion-compensated residual and is encoded together with the desired residual data, causing an increase in the coded bitrate. Camera noise can therefore significantly reduce the compression efficiency of an encoder. By filtering the input video sequence prior to encoding it may be possible to reduce camera noise (and hence improve compression efficiency). The filter parameters should be chosen with care, to avoid filtering out useful features of the video sequence.

Another phenomenon that can reduce compression efficiency is *camera shake*, small movements of the camera between successive frames, characteristic of a hand-held or poorly stabilised camera. These are 'seen' by the encoder as global motion between frames. Motion compensation may partly correct the motion but block-based motion estimation algorithms are not usually capable of correcting fully for camera shake and the result is an increase in residual energy and a drop in compression performance. Many consumer and professional camcorders incorporate image stabilisation systems that attempt to compensate automatically for small camera movements using mechanical and/or image processing methods. As well as improving the appearance of the captured video sequence, this has the effect of improving compression performance if the material is coded using motion compensation.

7.3.3 Post-processing

Video compression algorithms that incorporate quantisation (such as the core algorithms of MPEG-4 Visual and H.264) are inherently lossy, i.e. the decoded video frames are not identical to the original. The goal of any practical CODEC is to minimise distortion and maximise compression efficiency. It is often possible to reduce the actual or apparent distortion in the decoded video sequence by processing (filtering) the decoded frames. If the filtered decoded frames are then used for compensation, the filtering process can have the added benefit of improving motion-compensated prediction and hence compression efficiency.

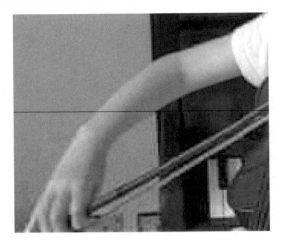

Figure 7.19   Distortion introduced by MPEG-4 Visual encoding (lower half)

Figure 7.20   Distortion introduced by H.264 encoding (lower half)

Block transform-based CODECs introduce characteristic types of distortion into the decoded video data. The lower half of Figure 7.19 shows typical distortion in a frame encoded and decoded using MPEG-4 Simple Profile (the upper half is the original, uncompressed frame). This example shows 'blocking' distortion (caused by mismatches at the boundaries of reconstructed 8 × 8 blocks) and 'ringing' distortion (faint patterns along the edges of objects, caused by the 'break through' of DCT basis patterns). Blocking is probably the most visually obvious (and therefore the most important) type of distortion introduced by video compression. Figure 7.20 (lower half) shows the result of encoding and decoding using H.264 without loop filtering. The smaller transform size in H.264 (4 × 4 rather than 8 × 8 samples) means that the blocking artefacts are correspondingly smaller, but are still obvious[1].

---

[1] The compressed halves of each of these figures were encoded at different bitrates.

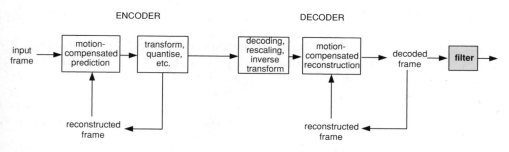

Figure 7.21 Post-filter implementation

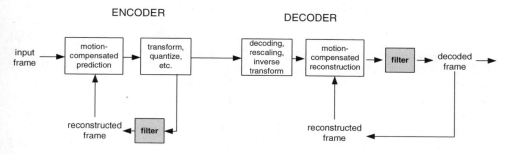

Figure 7.22 Loop filter implementation

Filters to reduce blocking (de-blocking) and/or ringing effects (de-ringing) are widely used in practical CODECs. Many filter designs have been proposed and implemented, ranging from relatively simple algorithms to iterative algorithms that are many times more complex than the encoding and decoding algorithms themselves [31–34]. The goal of a de-blocking or de-ringing filter is to minimise the effect of blocking or ringing distortion whilst preserving important features of the image. MPEG-4 Visual describes a deblocking filter and a deringing filter: these are 'informative' parts of the standard and are therefore optional. Both filters are designed to be placed at the output of the decoder (Figure 7.21). With this type of *post-filter*, unfiltered decoded frames are used as the reference for motion-compensated reconstruction of further frames. This means that the filters improve visual quality at the decoder but have no effect on the encoding and decoding processes.

It may be advantageous to place the filter inside the encoding and decoding 'loops' (Figure 7.22). At the decoder, the filtered decoded frame is stored for further motion-compensated reconstruction. In order to ensure that the encoder uses an identical reference frame, the same filter is applied to reconstructed frames in the encoder and the encoder uses the filtered frame as a reference for further motion estimation and compensation. If the quality of the filtered frame is better than that of an unfiltered decoded frame, then it will provide a better match for further encoded frames, resulting in a smaller residual after motion compensation and hence improved compression efficiency. H.264 makes use of this type of *loop filter* (see Chapter 6 for details of the filter algorithm). One disadvantage of incorporating the filter into the loop is that it must be specified in the standard (so that any decoder can successfully repeat the filtering process) and there is therefore limited scope for innovative filter designs.

7.4 PERFORMANCE

In this section we compare the performance of selected profiles of MPEG-4 Visual and H.264. It should be emphasised that what is considered to be 'acceptable' performance depends very much on the target application and on the type of video material that is encoded. Further, coding performance is strongly influenced by encoder decisions that are left to the discretion of the designer (e.g. motion estimation algorithm, rate control method, etc.) and so the performance achieved by a commercial CODEC may vary considerably from the examples reported here.

7.4.1 Criteria

Video CODEC performance can be considered as a tradeoff between three variables, quality, compressed bit rate and computational cost. 'Quality' can mean either subjective or objective measured video quality (see Chapter 2). Compressed bit rate is the rate (in bits per second) required to transmit a coded video sequence and computational cost refers to the processing 'power' required to code the video sequence. If video is encoded in real time, then the computational cost must be low enough to ensure encoding of at least n frames per second (where n is the target number of frames per second); if video is encoded 'offline', i.e. not in real time, then the computational cost per frame determines the total coding time of a video sequence.

The *rate–distortion* performance of a video CODEC describes the tradeoff between two of these variables, quality and bit rate. Plotting mean PSNR against coded bit rate produces a characteristic rate–distortion curve (Figure 7.23). As the bit rate is reduced, quality (as measured by PSNR) drops at an increasing rate. Plotting rate–distortion curves for identical source material (i.e. the same resolution, frame rate and content) is a widely accepted method of comparing the performance of two video CODECs. As Figure 7.23 indicates, 'better' rate–distortion performance is demonstrated by moving the graph up and to the left.

Comparing and evaluating competing video CODECs is a difficult problem. Desirable properties of a video CODEC include good rate–distortion performance and low (or acceptable) computational complexity. When comparing CODECs, it is important to use common

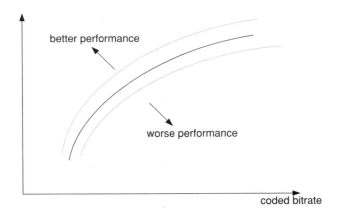

Figure 7.23   Example of a rate–distortion curve

test conditions where possible. For example, different video sequences can lead to dramatic differences in rate–distortion performance (i.e. some video sequences are 'easier' to code than others) and computational performance (especially if video processing is carried out in software). Certain coding artefacts (e.g. blocking, ringing) may be more visible in some decoded sequences than others. For example, blocking distortion is particularly visible in larger areas of continuously-varying tone in an image and blurring of features (for example due to a crude deblocking filter) is especially obvious in detailed areas of an image.

7.4.2 Subjective Performance

In this section we examine the subjective quality of video sequences after encoding and decoding. The 'Office' sequence (Figure 7.24) contains 200 frames, each captured in 4:2:0 CIF format (see Chapter 2 for details of this format). The 'Office' sequence was shot from a fixed camera position and the only movement is due to the two women. In contrast, the 'Grasses' sequence (Figure 7.25), also consisting of 200 CIF frames, was shot with a hand-held camera and contains rapid, complex movement of grass stems. This type of sequence is particularly difficult to encode due to the high detail and complex movement, since it is difficult to find accurate matches during motion estimation.

Each sequence was encoded using three CODECs, an MPEG-2 Video CODEC, an MPEG-4 Simple Profile CODEC and the H.264 Reference Model CODEC (operating in Baseline Profile mode, using only one reference picture for motion compensation). In each case, the first frame was encoded as an I-picture. The remaining frames were encoded as

Figure 7.24   Office: original frame

Figure 7.25   Grasses: original frame

P-pictures using the MPEG-4 and H.264 CODECs and as a mixture of B- and P-pictures with the MPEG-2 CODEC (with the sequence BBPBBP. . .). The 'Office' sequence was encoded at a mean bitrate of 150 kbps with all three CODECs and the 'Grasses' sequence at a mean bitrate of 900 kbps.

The decoded quality varies significantly between the three CODECs. A close-up of a frame from the 'Office' sequence after encoding and decoding with MPEG-2 (Figure 7.26) shows considerable blocking distortion and loss of detail. The MPEG-4 Simple Profile frame (Figure 7.27) is noticeably better but there is still evidence of blocking and ringing distortion. The H.264 frame (Figure 7.28) is the best of the three and at first sight there is little difference between this and the original frame (Figure 7.24). Visually important features such as the woman's face and smooth areas of continuous tone variation have been preserved but fine texture (such as the wood grain on the table and the texture of the wall) has been lost.

The results for the 'Grasses' sequence are less clear-cut. At 900 kbps, all three decoded sequences are clearly distorted. The MPEG-2 sequence (a close-up of one frame is shown in Figure 7.29) has the most obvious blocking distortion but blocking distortion is also clearly visible in the MPEG-4 Simple Profile sequence (Figure 7.30). The H.264 sequence (Figure 7.31) does not show obvious block boundaries but the image is rather blurred due to the deblocking filter. Played back at the full 25 fps frame rate, the H.264 sequence looks better than the other two but the performance improvement is not as clear as for the 'Office' sequence.

These examples highlight the way CODEC performance can change depending on the video sequence content. H.264 and MPEG-4 SP perform well at a relatively low bitrate (150 kbps) when encoding the 'Office' sequence; both perform significantly worse at a higher bitrate (900 kbps) when encoding the more complex 'Grasses' sequence.

Figure 7.26 Office: encoded and decoded, MPEG-2 Video (close-up)

Figure 7.27 Office: encoded and decoded, MPEG-4 Simple Profile (close-up)

Figure 7.28 Office: encoded and decoded, H.264 Baseline Profile (close-up)

Figure 7.29 Grasses: encoded and decoded, MPEG-2 Video (close-up)

Figure 7.30 Grasses: encoded and decoded, MPEG-4 Simple Profile (close-up)

Figure 7.31 Grasses: encoded and decoded, H.264 Baseline Profile (close-up)

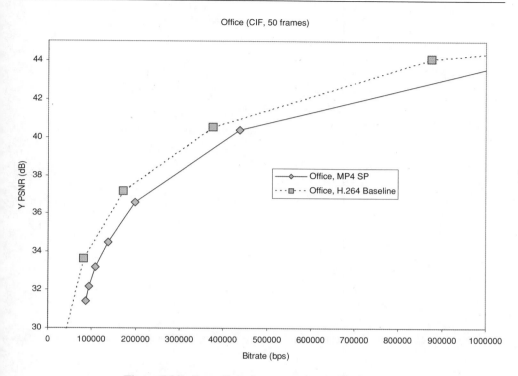

Figure 7.32   Rate–distortion comparison: 'Office', CIF

7.4.3 Rate–distortion Performance

Measuring bitrate and PSNR at a range of quantiser settings provides an estimate of compression performance that is numerically more accurate (but less closely related to visual perception) than subjective comparisons. Some comparisons of MPEG4 and H.264 performance are presented here.

Figure 7.32 and Figure 7.33 compare the performance of MPEG4 (Simple Profile) and H.264 (Baseline Profile, 1 reference frame) for the 'Office' and 'Grasses' sequences. Note that 'Office' is easier to compress than 'Grasses' (see above) and at a given bitrate, the PSNR of 'Office' is significantly higher than that of 'Grasses'. H.264 out-performs MPEG4 compression at all of the tested bit rates, but the rate-distortion gain is more noticeable for the 'Office' sequence.

The rate–distortion performance of the popular 'Carphone' test sequence is plotted in Figure 7.34. This sequence contains moderate motion and in this case the source is QCIF resolution at 30 frames per second. Four sets of results are compared, two from MPEG-4 and two from H.264. The first two are MPEG-4 Simple Profile (first frame coded as an I-picture, subsequent frames coded as P-pictures) and MPEG-4 Advanced Simple Profile (using two B-pictures between successive P-pictures, no other ASP tools used). The second pair are H.264 Baseline (first frame coded as an I-slice, subsequent frames coded as P-slices, one reference frame used for inter prediction, UVLC/CAVLC entropy coding) and H.264 Main Profile (first frame coded as an I-slice, subsequent frames coded as P-slices, five reference frames used for inter prediction, CABAC entropy coding).

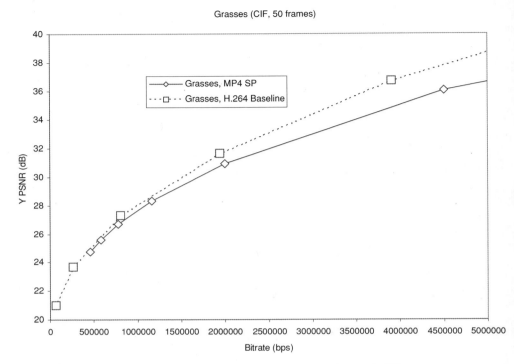

Figure 7.33   Rate–distortion comparison: 'Grasses', CIF

MPEG-4 ASP performs slightly better than MPEG-4 SP at higher bitrates but performs worse at lower bitrates (in this case). It is interesting to note that other ASP tools (quarter-pel MC and alternate quantiser) produced poorer performance in this test. H.264 Baseline outperforms both MPEG-4 profiles at all bitrates and CABAC and multiple reference frames provide a further performance gain. For example, at a PSNR of 35 dB, MPEG-4 SP produces a coded bitrate of around 125 kbps, ASP reduces the bitrate to around 120 kbps, H.264 Baseline (with one reference frame) produces a bitrate of around 80 kbps and H.264 with CABAC and five reference frames gives a bitrate of less than 70 kbps.

The results for the 'Carphone' sequence show a more convincing performance gain from H.264 than the results for 'Grasses' and 'Office'. This is perhaps because 'Carphone' is a professionally-captured sequence with high image fidelity whereas the other two sequences were captured from a high-end consumer video camera and have more noise in the original images.

Other Performance Studies

The compression performance of MPEG-4 Simple and Advanced Simple Profiles are compared in [35]. In this paper the Advanced Simple tools are found to improve the compression performance significantly, producing a coded bitrate around 30–40% smaller for the same video quality, at the expense of increased encoder complexity. Most of the performance gain

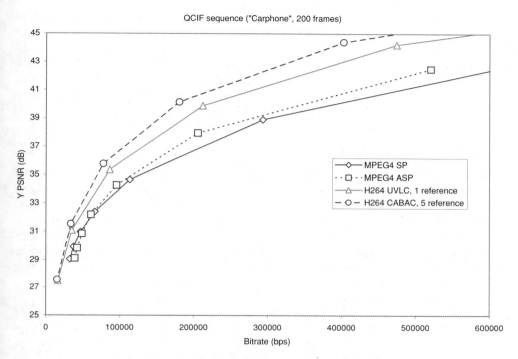

Figure 7.34 Rate–distortion comparison: 'Carphone', QCIF

is due to the use of B-pictures (which require extra storage and encoding delay), quarter-pixel motion compensation and rate–distortion optimised mode selection (i.e. choosing the encoding mode of each macroblock to maximise compression performance).

Reference [36] evaluates the performance of H.264 (an older version of the test model) and compares it with H.263++ (H.263 with optional modes to improve coding performance). According to the results presented, H.264 (H.26L) consistently out-performs H.263++. The authors study the contribution of some of the optional features of H.264 and conclude that CABAC provides a consistent coding gain compared with VLCs, reducing the coded bit rate by an average of 7.7% for the same objective quality (although the version of the test model did not include the more efficient context-adaptive VLCs for coefficient encoding). Using multiple reference frames for motion-compensated prediction provides better performance than a single reference frame, although the gains are slightly less obvious (a mean gain of 5.7% over the single reference frame case). Small motion compensation block sizes (down to 4 × 4) provided a clear performance gain over a single vector per macroblock (a mean bit rate saving of 16.4%), although most of the gain (13.7%) was achieved for a minimum block size of 8 × 8.

H.264 is compared with both H.263++ and MPEG-4 Visual Advanced Simple Profile in [37]. The authors compared the luminance PSNR (see Chapter 2) of each CODEC under the same test conditions. The three CODECs were chosen to represent the most highly-optimised versions of each standard available at the time. Test Model 8 of H.26L (an earlier version of H.264) out-performed H.263++ and MPEG-4 ASP by an average of 3.0 dB and 2.0 dB respectively. The H.26L CODEC achieved roughly the same performance at a coded bitrate of 32 kbit/s as the other two CODECs at a bitrate of 64 kbit/s (QCIF video, 10 frames per second), i.e. in this test H.26L produced the same decoded quality at around half the bitrate of MPEG-4 ASP and H.263++. At higher bitrates (512 kbps and above) the gain was still significant (but not so large). An overview of rate-constrained encoder control and a comparison of H.264 performance with H.263, MPEG-2 Video and MPEG-4 Visual is given in [38].

7.4.4 Computational Performance

MPEG-4 Visual and (to a lesser extent) H.264 provide a range of optional coding modes that have the potential to improve compression performance. For example, MPEG-4's Advanced Simple Profile is designed to offer greater compression efficiency than the popular Simple Profile (see Chapter 5); the Main Profile of H.264 is capable of providing better compression efficiency than the Baseline Profile (see Chapter 6). Within a specific Profile, a designer or user of a CODEC can choose whether or not to enable certain coding features. A Main Profile H.264 decoder should support both Context Adaptive VLCs (CAVLC) and arithmetic coding (CABAC) but the encoder has the choice of which mode to use in a particular application.

Improved coding efficiency often comes at the cost of higher computational complexity. The situation is complicated by the fact that the computational cost and coding benefit of a particular mode or feature can depend very much on the type of source material. In a practical application, the choice of possible coding modes may depend on the limitations of the processing platform and it may be necessary to choose encoding parameters to suit the source material and available processing resources.

Example

The first 25 frames of the 'Violin' sequence (QCIF, 25 frames per second, see Figure 7.18) were encoded using the H.264 test model software (version JM4.0) with a fixed quantiser parameter of 36. The sequence was encoded with a range of coding parameters to investigate the effect of each on compression performance and coding time. Two reference configurations were used as follows.

Basic configuration: CAVLC entropy coding, no B-pictures, loop filter enabled, rate–distortion optimisation disabled, one reference frame for motion compensation, all block sizes (down to 4 × 4) available.

Advanced configuration: CABAC entropy coding, every 2nd picture coded as a B-picture, loop filter enabled, rate–distortion optimisation enabled, five reference frames, all block sizes available.

The 'basic' configuration represents a suitable set-up for a low complexity, real-time CODEC whilst the 'advanced' configuration might be suitable for a high-complexity, high-efficiency CODEC. Table 7.3 summarises the results. The luminance PSNR (objective quality) of each sequence is almost identical and the differences in performance are apparent in the coded bitrate and encoding time.

The 'basic' configuration takes 40 seconds to encode the sequence and produces a bitrate of 46 kbps (excluding the bits produced by the first I-slice). Using only 8 × 8 or larger motion compensation block sizes reduces coding time (by c. 6 seconds) but increases the coded bitrate, as

Table 7.3 Computational performance of H.264 optional modes: violin, QCIF, 25 frames

Configuration	Average luminance PSNR (dB)	Coded bitrate (P/B slices) (kbps)	Encoding time (seconds)
Basic	29.06	45.9	40.4
Basic + min. block size of 8 × 8	29.0	46.6	33.9
Basic + 5 reference frames	29.12	46.2	157.2
Basic + rate-distortion optimisation	29.18	44.6	60.5
Basic + every 2nd picture coded as a B-picture	29.19	42.2	55.7
Basic + CABAC	29.06	44.0	40.5
Advanced	29.57	38.2	180
Advanced (only one reference frame)	29.42	38.8	77

might be expected. Using multiple reference frames (five in this case) increases coding time (by almost four times) but results in an *increase* in coded bitrate. Adding rate–distortion optimisation (in which the encoder repeatedly codes each macroblock in different ways in order to find the best coding parameters) reduces the bitrate at the expense of a 50% increase in coding time. B-pictures provide a compression gain at the expense of increased coding time (nearly 50%); CABAC gives a compression improvement and does not increase coding time.

The 'advanced' configuration takes over four times longer than the 'basic' configuration to encode but produces a bitrate 17% smaller than the basic configuration. By using only one reference frame, the coding time is reduced significantly at the expense of a slight drop in compression efficiency.

These results show that, for this sequence and this encoder at least, the most useful performance optimisations (in terms of coding efficiency improvement and computational complexity) are CABAC and B-pictures. These give a respectable improvement in compression without a high computational penalty. Conversely, multiple reference frames make only a slight improvement (and then only in conjunction with certain other modes, notably rate-distortion optimised encoding) and are computationally expensive. It is worth noting, however, (i) that different outcomes would be expected with other types of source material (for example, see [36]) and (ii) that the reference model encoder is not optimised for computational efficiency.

7.4.5 Performance Optimisation

Achieving the optimum balance between compression and decoded quality is a difficult and complex challenge. Setting encoding parameters at the start of a video sequence and leaving them unchanged throughout the sequence is unlikely to produce optimum rate–distortion performance since the encoder faces a number of inter-related choices when coding each macroblock. For example, the encoder may select a motion vector for an inter-coded MB that minimises the energy in the motion-compensated residual. However, this is not necessarily the best choice because larger MVs generally require more bits to encode and the optimum choice of MV is the one that minimises the total number of bits in the coded MB (including header, MV and coefficients). Thus finding the optimal choice of parameters (such as MV, quantisation parameter, etc.) may require the encoder to code the MB repeatedly before selecting the combination of parameters that minimise the coded size of the MB. Further, the choice of parameters for MB1 affects the coding performance of MB2 since, for example, the coding modes of MB2 (e.g. MV, intra prediction mode, etc.) may be differentially encoded from the coding modes of MB1.

Achieving near-optimum rate–distortion performance can be a very complex problem indeed, many times more complex than the video coding process itself. In a practical CODEC, the choice of optimisation strategy depends on the available processing power and acceptable coding latency. So-called 'two-pass' encoding is widely used in offline encoding, in which each frame is processed once to generate sequence statistics which then influence the coding strategy in the second coding pass (often together with a rate control algorithm to achieve a target bit rate or file size).

Many alternative rate–distortion optimisation strategies have been proposed (such as those based on Lagrangian optimisation) and a useful review can be found in [6]. Rate–distortion optimisation should not be considered in isolation from computational performance. In fact, video CODEC optimisation is (a least) a three-variable problem since rate, distortion and computational complexity are all inter-related. For example, rate–distortion optimised mode decisions are achieved at the expense of increased complexity, 'fast' motion estimation algorithms often achieve low complexity at the expense of motion estimation (and hence coding) performance, and so on. Coding performance and computational performance can be traded against each other. For example, a real-time coding application for a hand-held device may be designed with minimal processing load at the expense of poor rate–distortion performance, whilst an application for offline encoding of broadcast video data may be designed to give good rate–distortion performance, since processing time is not an important issue but encoded quality is critical.

7.5 RATE CONTROL

The MPEG-4 Visual and H.264 standards require each video frame or object to be processed in units of a macroblock. If the control parameters of a video encoder are kept constant (e.g. motion estimation search area, quantisation step size, etc.), then the number of coded bits produced for each macroblock will change depending on the content of the video frame, causing the bit rate of the encoder output (measured in bits per coded frame or bits per second of video) to vary. Typically, an encoder with constant parameters will produce more bits when there is high motion and/or detail in the input sequence and fewer bits when there is low motion and/or detail. Figure 7.35 shows an example of the variation in output bitrate produced by coding the Office sequence (25 frames per second) using an MPEG-4 Simple Profile encoder, with a fixed quantiser step size of 12. The first frame is coded as an I-VOP (and produces a large number of bits because there is no temporal prediction) and successive frames are coded as P-VOPs. The number of bits per coded P-VOP varies between 1300 and 9000 (equivalent to a bitrate of 32–225 kbits per second).

This variation in bitrate can be a problem for many practical delivery and storage mechanisms. For example, a constant bitrate channel (such as a circuit-switched channel) cannot transport a variable-bitrate data stream. A packet-switched network can support varying throughput rates but the mean throughput at any point in time is limited by factors such as link rates and congestion. In these cases it is necessary to adapt or control the bitrate produced by a video encoder to match the available bitrate of the transmission mechanism. CD-ROM

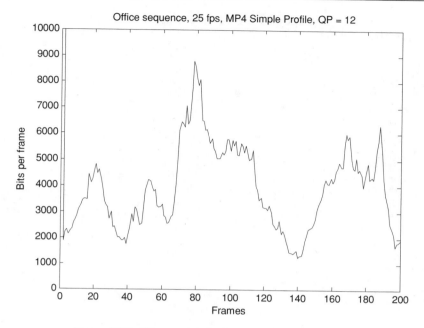

Figure 7.35 Bit rate variation (MPEG-4 Simple Profile)

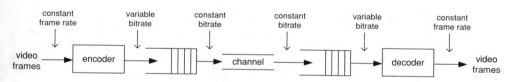

Figure 7.36 Encoder output and decoder input buffers

and DVD media have a fixed storage capacity and it is necessary to control the rate of an encoded video sequence (for example, a movie stored on DVD-Video) to fit the capacity of the medium.

The variable data rate produced by an encoder can be 'smoothed' by buffering the encoded data prior to transmission. Figure 7.36 shows a typical arrangement, in which the variable bitrate output of the encoder is passed to a 'First In/First Out' (FIFO) buffer. This buffer is emptied at a constant bitrate that is matched to the channel capacity. Another FIFO is placed at the input to the decoder and is filled at the channel bitrate and emptied by the decoder at a variable bitrate (since the decoder extracts P bits to decode each frame and P varies).

Example

The 'Office' sequence is coded using MPEG-4 Simple Profile with a fixed $QP = 12$ to produce the variable bitrate plotted in Figure 7.35. The encoder output is buffered prior to transmission over a 100 kbps constant bitrate channel. The video frame rate is 25 fps and so the channel transmits

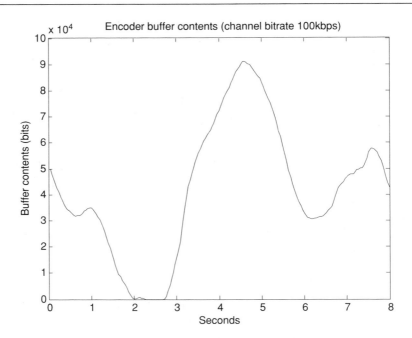

Figure 7.37   Buffer example (encoder; channel bitrate 100 kbps)

4 kbits (and hence removes 4 kbits from the buffer) in every frame period. Figure 7.37 plots the contents of the encoder buffer (y-axis) against elapsed time (x-axis). The first I-VOP generates over 50 kbits and subsequent P-VOPs in the early part of the sequence produce relatively few bits and so the buffer contents drop for the first 2 seconds as the channel bitrate exceeds the encoded bitrate. At around 3 seconds the encoded bitrate starts to exceed the channel bitrate and the buffer fills up.

Figure 7.38 shows the state of the decoder buffer, filled at a rate of 100 kbps (4 kbits per frame) and emptied as the decoder extracts each frame. It takes half a second before the first complete coded frame (54 kbits) is received. From this point onwards, the decoder is able to extract and decode frames at the correct rate (25 frames per second) until around 4 seconds have elapsed. At this point, the decoder buffer is emptied and the decoder 'stalls' (i.e. it has to slow down or pause decoding until enough data are available in the buffer). Decoding picks up again after around 5.5 seconds.

If the decoder stalls in this way it is a problem for video playback because the video clip 'freezes' until enough data available to continue. The problem can be partially solved by adding a deliberate delay at the decoder. For example, Figure 7.39 shows the results if the decoder waits for 1 second before it starts decoding. Delaying decoding of the first frame allows the buffer contents to reach a higher level before decoding starts and in this case the contents never drop to zero and so playback can proceed smoothly[2].

---

[2] Varying throughput rates from the *channel* can also be handled using a decoder buffer. For example, a widely-used technique for video streaming over IP networks is for the decoder to buffer a few seconds of coded data before commencing decoding. If data throughput drops temporarily (for example due to network congestion) then decoding can continue as long as data remain in the buffer.

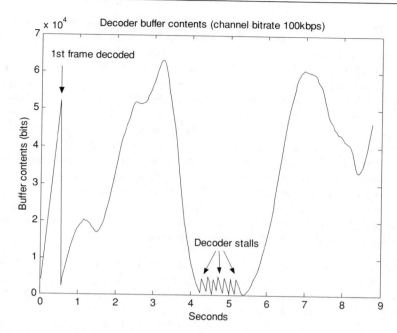

Figure 7.38   Buffer example (decoder; channel bitrate 100 kbps)

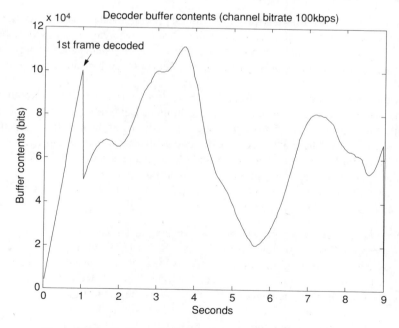

Figure 7.39   Buffer example (decoder; channel bitrate 100 kbps)

These examples show that a variable coded bitrate can be adapted to a constant bitrate delivery medium using encoder and decoder buffers. However, this adaptation comes at a cost of buffer storage space and delay and (as the examples demonstrate) the wider the bitrate variation, the larger the buffer size and decoding delay. Furthermore, it is not possible to cope with an arbitrary variation in bitrate using this method, unless the buffer sizes and decoding delay are set at impractically high levels. It is usually necessary to implement a feedback mechanism to control the encoder output bitrate in order to prevent the buffers from over- or under-flowing.

Rate control involves modifying the encoding parameters in order to maintain a target output bitrate. The most obvious parameter to vary is the quantiser parameter or step size (QP) since increasing QP reduces coded bitrate (at the expense of lower decoded quality) and vice versa. A common approach to rate control is to modify QP during encoding in order to (a) maintain a target bitrate (or mean bitrate) and (b) minimise distortion in the decoded sequence. Optimising the tradeoff between bitrate and quality is a challenging task and many different approaches and algorithms have been proposed and implemented. The choice of rate control algorithm depends on the nature of the video application, for example:

(a) Offline encoding of stored video for storage on a DVD. Processing time is not a particular constraint and so a complex algorithm can be employed. The goal is to 'fit' a compressed video sequence into the available storage capacity whilst maximising image quality and ensuring that the decoder buffer of a DVD player does not overflow or underflow during decoding. Two-pass encoding (in which the encoder collects statistics about the video sequence in a first pass and then carries out encoding in a second pass) is a good option in this case.

(b) Encoding of live video for broadcast. A broadcast programme has one encoder and multiple decoders; decoder processing and buffering is limited whereas encoding may be carried out in expensive, fast hardware. A delay of a few seconds is usually acceptable and so there is scope for a medium-complexity rate control algorithm, perhaps incorporating two-pass encoding of each frame.

(c) Encoding for two-way videoconferencing. Each terminal has to carry out both encoding and decoding and processing power may be limited. Delay must be kept to a minimum (ideally less than around 0.5 seconds from frame capture at the encoder to display at the decoder). In this scenario a low-complexity rate control algorithm is appropriate. Encoder and decoder buffering should be minimised (in order to keep the delay small) and so the encoder must tightly control output rate. This in turn may cause decoded video quality to vary significantly, for example it may drop significantly when there is an increase in movement or detail in the video scene.

Recommendation H.264 does not (at present) specify or suggest a rate control algorithm (however, a proposal for H.264 rate control is described in [39]). MPEG-4 Visual describes a possible rate control algorithm in an Informative Annex [40] (i.e. use of the algorithm is not mandatory). This algorithm, known as the Scalable Rate Control (SRC) scheme, is appropriate for a single video object (a rectangular V.O. that covers the entire frame) and a range of bit rates and spatial/temporal resolutions. The SRC attempts to achieve a target bit rate over a certain number of frames (a 'segment' of frames, usually starting with an I-VOP) and assumes the following model for the encoder rate R:

$$R = \frac{X_1 S}{Q} + \frac{X_2 S}{Q^2} \qquad (7.10)$$

where Q is the quantiser step size, S is the mean absolute difference of the residual frame after motion compensation (a measure of frame complexity) and X_1, X_2 are model parameters. Rate control consists of the following steps which are carried out after motion compensation and before encoding of each frame i:

1. Calculate a target bit rate R_i, based on the number of frames in the segment, the number of bits that are available for the remainder of the segment, the maximum acceptable buffer contents and the estimated complexity of frame i. (The maximum buffer size affects the latency from encoder input to decoder output. If the previous frame was complex, it is assumed that the next frame will be complex and should therefore be allocated a suitable number of bits: the algorithm attempts to balance this requirement against the limit on the total number of bits for the segment.)
2. Compute the quantiser step size Q_i (to be applied to the whole frame). Calculate S for the complete residual frame and solve equation (7.10) to find Q.
3. Encode the frame.
4. Update the model parameters X_1, X_2 based on the actual number of bits generated for frame i.

The SRC algorithm aims to achieve a target bit rate across a segment of frames (rather than a sequence of arbitrary length) and does not modulate the quantiser step size within a coded frame, giving a uniform visual appearance within each frame but making it difficult to maintain a small buffer size and hence a low delay. An extension to the SRC supports modulation of the quantiser step size at the macroblock level and is suitable for low-delay applications that require 'tight' rate control. The macroblock-level algorithm is based on a model for the number of bits B_i required to encode macroblock i, equation (7.11):

$$B_i = A \left(K \frac{\sigma_i^2}{Q_i^2} + C \right) \qquad (7.11)$$

where A is the number of pixels in a macroblock, σ_i is the standard deviation of luminance and chrominance in the residual macroblock (i.e. a measure of variation within the macroblock), Q_i is the quantisation step size and K, C are constant model parameters. The following steps are carried out for each macroblock i:

1. Measure σ_i.
2. Calculate Q_i based on B, K, C, σ_i and a macroblock weight α_i.
3. Encode the macroblock.
4. Update the model parameters K and C based on the actual number of coded bits produced for the macroblock.

The weight α_i controls the 'importance' of macroblock i to the subjective appearance of the image and a low value of α_i means that the current macroblock is likely to be highly quantised. These weights may be selected to minimise changes in Q_i at lower bit rates since each change involves sending a modified quantisation parameter DQUANT which means encoding an extra five bits per macroblock. It is important to minimise the number of changes to Q_i during encoding of a frame at low bit rates because the extra five bits in a macroblock may become significant; at higher bit rates, this DQUANT overhead is less important and Q may change more frequently without significant penalty. This rate control method is effective at maintaining good visual quality with a small encoder output buffer, keeping coding delay to a minimum (important for low-delay applications such as scenario (c) described above).

Further information on some of the many alternative strategies for rate control can be found in [41].

7.6 TRANSPORT AND STORAGE

A video CODEC is rarely used in isolation; instead, it is part of a communication system that involves coding video, audio and related information, combining the coded data and storing and/or transmitting the combined stream. There are many different options for combining (multiplexing), transporting and storing coded multimedia data and it has become clear in recent years that no single transport solution fits every application scenario.

7.6.1 Transport Mechanisms

Neither MPEG-4 nor H.264 define a mandatory transport mechanism for coded visual data. However, there are a number of possible transport solutions depending on the method of transmission, including the following.

MPEG-2 Systems: Part 1 of the MPEG-2 standard [42] defines two methods of multiplexing audio, video and associated information into streams suitable for transmission (Program Streams or Transport Streams). Each data source or *elementary stream* (e.g. a coded video or audio sequence) is packetised into Packetised Elementary Stream (PES) packets and PES packets from the different elementary streams are multiplexed together to form a Program Stream (typically carrying a single set of audio/visual data such as a single TV channel) or a Transport Stream (which may contain multiple channels) (Figure 7.40). The Transport Stream adds both Reed–Solomon and convolutional error control coding and so provides protection from transmission errors. Timing and synchronisation is supported by a system of clock references and time stamps in the sequence of packets. An MPEG-4 Visual stream may be carried as an elementary stream within an MPEG-2 Program or Transport Stream. Carriage of an MPEG-4 Part 10/H.264 stream over MPEG-2 Systems is covered by Amendment 3 to MPEG-2 Systems, currently undergoing standardisation.

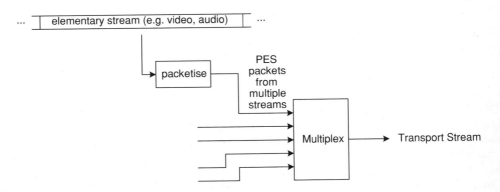

Figure 7.40   MPEG-2 Transport Stream

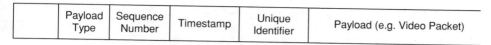

	Payload Type	Sequence Number	Timestamp	Unique Identifier	Payload (e.g. Video Packet)

Figure 7.41 RTP packet structure (simplified)

Real-Time Protocol: RTP [43] is a packetisation protocol that may be used in conjunction with the User Datagram Protocol (UDP) to transport real-time multimedia data across networks that use the Internet Protocol (IP). UDP is preferable to the Transmission Control Protocol (TCP) for real-time applications because it offers low-latency transport across IP networks. However, it has no mechanisms for packet loss recovery or synchronisation. RTP defines a packet structure for real-time data (Figure 7.41) that includes a type identifier (to signal the type of CODEC used to generate the data), a sequence number (essential for reordering packets that are received out of order) and a time stamp (necessary to determine the correct presentation time for the decoded data). Transporting a coded audio-visual stream via RTP involves packetising each elementary stream into a series of RTP packets, interleaving these and transmitting them across an IP network (using UDP as the basic transport protocol). RTP *payload formats* are defined for various standard video and audio CODECs, including MPEG-4 Visual and H.264. The NAL structure of H.264 (see Chapter 6) has been designed with efficient packetisation in mind, since each NAL unit can be placed in its own RTP packet.

MPEG-4 Part 6 defines an optional session protocol, the Delivery Multimedia Integration Framework, that supports session management of MPEG-4 data streams (e.g. visual and audio) across a variety of network transport protocols. The FlexMux tool (part of MPEG-4 Systems) provides a flexible, low-overhead mechanism for multiplexing together separate Elementary Streams into a single, interleaved stream. This may be useful for multiplexing separate audio-visual objects prior to packetising into MPEG-2 PES packets, for example.

7.6.2 File Formats

Earlier video coding standards such as MPEG-1, MPEG-2 and H.263 did not explicitly define a format for storing compressed audiovisual data in a file. It is common for single compressed video sequences to be stored in files, simply by mapping the encoded stream to a sequence of bytes in a file, and in fact this is a commonly used mechanism for exchanging test bitstreams. However, storing and playing back combined audio-visual data requires a more sophisticated file structure, especially when, for example, the stored data is to be streamed across a network or when the file is required to store multiple audio-visual objects. The MPEG-4 File Format and AVC File Format (which will both be standardised as Parts of MPEG-4) are designed to store MPEG-4 Audio-Visual and H.264 Video data respectively. Both formats are derived from the ISO Base Media File Format, which in turn is based on Apple Computer's QuickTime format.

In the ISO Media File Format, a coded stream (for example an H.264 video sequence, an MPEG-4 Visual video object or an audio stream) is stored as a *track*, representing a sequence of coded data items (*samples*, e.g. a coded VOP or coded slice) with time stamps (Figure 7.42). The file formats deal with issues such as synchronisation between tracks, random access indices and carriage of the file on a network transport mechanism.

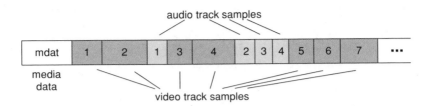

Figure 7.42   ISO Media File

7.6.3 Coding and Transport Issues

Many of the features and tools of the MPEG-4 Visual and H.264 standards are primarily aimed at improving compression efficiency. However, it has long been recognised that it is necessary to take into account practical transport issues in a video communication system and a number of tools in each standard are specifically designed to address these issues.

Scaling a delivered video stream to support decoders with different capabilities and/or delivery bitrates is addressed by both standards in different ways. MPEG-4 Visual includes a number of tools for scalable coding (see Chapter 5), in which a sequence or object is coded to produce a number of layers. Typically, these include a base layer (which may be decoded to obtain a 'basic' quality version of the sequence) and enhancement layer(s), each of which requires an increased transmission bitrate but which adds quality (e.g. image quality, spatial or temporal resolution) to the decoded sequence. H.264 takes a somewhat different approach. It does not support scalable coding but provides SI and SP slices (see Chapter 6) that enable a decoder to switch efficiently between multiple coded versions of a stream. This can be particularly useful when decoding video streamed across a variable-throughput network such as the Internet, since a decoder can dynamically select the highest-rate stream that can be delivered at a particular time.

Latency is a particular issue for two-way real time appliations such as videoconferencing. Tools such as B-pictures (coded frames that use motion-compensated prediction from earlier and later frames in temporal order) can improve compression efficiency but introduce a delay of several frame periods into the coding and decoding 'chain' which may be unacceptable for low-latency two way applications. Latency requirements also have an influence on rate control algorithms (see Section 7.5) since post-encoder and pre-decoder buffers (useful for smoothing out rate variations) increase latency.

Each standard includes a number of features to aid the handling of transmission errors. Bit errors are a characteristic of circuit-switched channels; packet-switched networks tend to suffer from packet losses (since a bit error in a packet typically results in the packet being dropped during transit). Errors can have a serious impact on decoded quality [44] because the effect of an error may propagate spatially (distorting an area within the current decoded frame) and temporally (propagating to successive decoded frames that are temporally predicted from the errored frame). Chapters 5 and 6 describe tools that are specifically intended to reduce the damage caused by errors, including data partitioning and independent slice decoding (designed to limit error propagation by localising the effect of an error), redundant slices (sending extra copies of coded data), variable-length codes that can be decoded in either direction (reducing the likelihood of a bit error 'knocking out' the remainder of a coded unit) and flexible ordering of macroblocks and slices (to make it easier for a decoder to conceal the effect of an error by interpolating from neighbouring error-free data).

7.7 CONCLUSIONS

Different choices during the design of a CODEC and different strategies for coding control can lead to significant variations in compression and computational performance between CODEC implementations. However, the best performance that may be achieved by a CODEC is limited by the available coding tools. The performance examples presented here and many other studies in the literature indicate that H.264 has the ability to out-perform MPEG-4 Visual convincingly (which in turn performs significantly better than MPEG-2). Performance is only one of many factors that influence whether a new technology is successful in the marketplace and in the final chapter we examine some of the other issues that are currently shaping the commercial market for video coding.

7.8 REFERENCES

1. ISO/IEC 14496-2, Coding of audio-visual objects – Part 2: Visual, 2001, Annex F.
2. S. Sun, D. Haynor and Y. Kim, Semiautomatic video object segmentation using VSnakes, *IEEE Trans. Circuits* Syst. Video Technol., **13** (1), January 2003.
3. C. Kim and J-N Hwang, Fast and automatic video object segmentation and tracking for content-based applications, *IEEE Trans. Circuits* Syst. Video Technol., **12** (2), February 2002.
4. J. Kim and T. Chen, A VLSI architecture for video-object segmentation, *IEEE Trans. Circuits* Syst. Video Technol., **13** (1), January 2003.
5. H. 264 reference model software version JM6.1b, http://bs.hhi.de/~suehring/tml/, March 2003.
6. G. Sullivan and T. Wiegand, Rate-distortion optimization for video compression, IEEE Signal Process. Mag., November 1998.
7. T. Koga, K. Iinuma *et al.*, Motion compensated interframe coding for video conference, *Proc. NTC*, November 1991.
8. J. R. Jain and A. K. Jain, Displacement measurement and its application in interframe image coding, *IEEE Trans.* Commun., **29**, December 1981.
9. M. Ghanbari, The cross-search algorithm for motion estimation, *IEEE Trans.* Commun., **38**, July 1990.
10. M. Gallant, G. Côté and F. Kossentini, An efficient computation-constrained block-based motion estimation algorithm for low bit rate video coding, *IEEE Trans. Image Processing*, **8** (12), December 1999.
11. P. Kuhn, G. Diebel, S. Hermann, A. Keil, H. Mooshofer, A. Kaup, R. Mayer and W. Stechele, Complexity and PSNR-Comparison of Several Fast Motion Estimation Algorithms for MPEG-4, *Proc. Applications of Digital Image Processing XXI*, San Diego, 21–24 July 1998; *SPIE*, **3460**, pp. 486–499.
12. R. Garg, C. Chung, D. Kim and Y. Kim, Boundary macroblock padding in MPEG-4 video decoding using a graphics coprocessor, *IEEE Trans. Circuits* Syst. Video Technol., **12** (8), August 2002.
13. H. Chang, Y-C Chang, Y-C Wang, W-M Chao and L-G Chen, VLSI Architecture design of MPEG-4 shape coding, *IEEE Trans. Circuits* Syst. Video Technol., **12** (9), September 2002.
14. W-H Chen, C. H. Smith and S. C. Fralick, A fast computational algorithm for the discrete cosine transform, *IEEE Trans.* Commun., **COM-25**(9), September 1977.
15. I. E. G. Richardson, *Video Codec Design*, Wiley, 2002.

16. J. R. Spanier, G. Keane, J. Hunter and R. Woods, Low power implementation of a discrete cosine transform IP core, *Proc. DATE-2000*, Paris, March 2000.

17. G. Aggarwal and D. Gajski, Exploring DCT Implementations, UC Irvine Tech Report TR-98-10, March 1998.

18. T-S Chang, C-S Kung and C-W Jen, A simple processor core design for DCT/IDCT, *IEEE Trans. CSVT*, **10** (3), April 2000.

19. A. Hallapuro and M. Karczewicz, Low complexity transform and quantisation – Part 1: Basic implementation, JVT document JVT-B038, February 2001.

20. L. Kerofsky, Matrix IDCT, JVT document JVT-E033, October 2002.

21. K. Takagi, A. Koike and S. Matsumoto, Padding method for arbitrarily-shaped region coding based on rate-distortion properties, *Trans. IEICE D-II*, pp 238–247, February 2001.

22. A. Kaup, Object-based texture coding of moving video in MPEG-4, *IEEE Trans. Circuits* Syst. Video Technol., **9** (1), February 1999.

23. O. Rioul and P. Duhamel, Fast algorithms for wavelet transform computation, Chapter 8 in *Time-frequency and Wavelets in Biomedical Engineering*, pp. 211–242, M. Akay (ed.), IEEE Press, 1997.

24. W. Jiang and A. Ortega, Lifting factorization based discrete wavelet transform architecture design, *IEEE Trans. Circuits* Syst. Video Technol., **11** (5), pp. 651–657, May 2001.

25. M. Ravasi, L. Tenze and M. Mattaveli, A scalable and programmable architecture for 2D DWT decoding, *IEEE Trans. Circuits* Syst. Video Technol., **12** (8), August 2002.

26. S. M. Lei and M-T Sun, An entropy coding system for digital HDTV applications, *IEEE Trans. CSVT*, **1** (1), March 1991.

27. H-C Chang, L-G Chen, Y-C Chang, and S-C Huang, A VLSI architecture design of VLC encoder for high data rate video/image coding, *1999 IEEE Int. Symp. Circuits and Systems* (ISCAS'99).

28. S. F. Chang and D. Messerschmitt, Designing high-throughput VLC decoder, Part I – concurrent VLSI architectures, *IEEE Trans. CSVT*, **2**(2), June 1992,

29. B-J Shieh, Y-S Lee and C-Y Lee, A high throughput memory-based VLC decoder with codeword boundary prediction, *IEEE Trans. CSVT*, **10**(8), December 2000.

30. D. Marpe, H. Schwarz and T. Wiegand, "Context-Based Adaptive Binary Arithmetic Coding in the H.264/AVC Video Compression Standard", IEEE Transactions on Circuits and Systems for Video Technology, to be published in 2003.

31. J. Chou, M. Crouse and K. Ramchandran, A simple algorithm for removing blocking artifacts in block transform coded images, *IEEE Signal Process. Lett.*, **5**, February 1998.

32. S. Hong, Y. Chan and W. Siu, A practical real-time post-processing technique for block effect elimination, *Proc. IEEE ICIP96*, Lausanne, September 1996.

33. T. Meier, K. Ngan and G. Crebbin, Reduction of coding artifacts at low bit rates, *Proc. SPIE Visual Communications and Image Processing*, San Jose, January 1998.

34. Y. Yang and N. Galatsanos, Removal of compression artifacts using projections onto convex sets and line modeling, *IEEE Trans.* Image Processing, **6**, October 1997.

35. ISO/IEC JTC1/SC29/WG11/M7227, Performance of MPEG-4 profiles used for streaming video and comparison with H.26L, Sydney, July 2001.

36. A. Joch and F. Kossentini, Performance analysis of H.26L coding features, ITU-T Q.6/SG16 VCEG-O042, Pattaya, November 2001.

37. P. Topiwala, G. Sullivan, A. Joch and F. Kossentini, Performance evaluation of H.26L TML8 vs. H.263++ and MPEG-4, ITU-T Q.6/SG16 VCEG-N18, September 2001.

38. T. Wiegand, H. Schwarz, A Joch, F. Kossentini and G.Sullivan, "Rate-Constrained Coder Control and Comparison of Video Coding Standards", IEEE Transactions on Circuits and Systems for Video Technology, to be published in 2003.

39. Z. Li, W. Gao et al, "Adaptive Rate Control with HRD Consideration", ISO/IEC JTC1/SC29/WG11 and ITU-T SG16 Q.6 Document JVT-H014, May 2003.

40. ISO/IEC 14496-2, Coding of audio-visual objects – Part 2: Visual, 2001, Annex L.

41. Y-S Saw, *Rate-Quality Optimized Video Coding*, Kluwer Academic Publishers, November 1998.
42. ISO/IEC 13818, Information technology: generic coding of moving pictures and associated audio information, 1995 (MPEG-2).
43. IETF RFC 1889, RTP: A transport protocol for real-time applications, January 1996.
44. A. H. Sadka, *Compressed Video Communications*, John Wiley & Sons, 2002.

8

Applications and Directions

8.1 INTRODUCTION

In addition to technical features and theoretical performance, there are many other important considerations to take into account when choosing a CODEC for a practical application. These include industry support for the CODEC (for example, whether there are multiple sources for technology), availability of development tools and costs (including development, integration and licensing costs). In this chapter we discuss the requirements of a selection of applications and consider practical issues that may affect the choice of CODEC. We list some of the available commercial MPEG-4 Visual and H.264 CODECs and report the claimed performance of these CODECs (where available). We review important commercial considerations (including the contentious topic of patent licensing) and end by making some predictions about the short- and long-term developments in CODEC research and standardisation.

8.2 APPLICATIONS

Table 8.1 lists a selection of applications for video coding technology together with important requirements for each application and suggested MPEG-4 Visual or H.264 Profiles.

None of the applications listed in Table 8.1 require object-based coding tools. As is clear from the list of commercially-available CODECs in Section 8.4 object-based coding is not widely supported in the marketplace and few, if any, 'real' commercial applications of object-based coding have emerged since the release of MPEG-4 Visual. The following anecdote is perhaps typical of the situation with regard to object-based coding, which has been described as 'a technology looking for an application'.

Some promising work on 'closed signing' was presented at a recent workshop on visual communications for deaf users ('Silent Progress', hosted by Bristol University, March 2003). Closed signing involves transmitting a separate channel as part of a digital television multi-plexing, incorporating a human sign language interpreter translating the content of the main channel. MPEG-4's object-based coding tools were used to code the signer as a video object,

H.264 and MPEG-4 Video Compression: Video Coding for Next-generation Multimedia.
Iain E. G. Richardson.   © 2003 John Wiley & Sons, Ltd. ISBN: 0-470-84837-5

Table 8.1 Application requirements

Application	Requirements	MPEG-4 Profiles*	H.264 Profiles
Broadcast television	Coding efficiency, reliability (over a 'controlled' distribution channel), interlace, low-complexity decoder	ASP	Main
Streaming video	Coding efficiency, reliability (over an 'uncontrolled' packet-based network), scalability	ARTS or FGS	Extended
Video storage and playback (e.g. DVD)	Coding efficiency, interlace, low-complexity decoder	ASP	Main
Videoconferencing	Coding efficiency, reliability, low latency, low-complexity encoder and decoder	SP	Baseline
Mobile video	Coding efficiency, reliability, low latency, low-complexity encoder and decoder, low power consumption	SP	Baseline
Studio distribution	Lossless or near-lossless, interlace, efficient transcoding	Studio	Main

*SP, ASP, ARTS, FGS, Studio: Simple, Advanced Simple, Advanced Real Time Simple, Fine Granular Scalability and Studio Profiles.

making it possible to superimpose the figure of the signer on the main TV programme display. However, the feedback from deaf users at the workshop indicated that the superimposed figure was distracting and hard to follow and the users much preferred to see the signer in a separate (rectangular!) window on the screen. It was concluded that the demonstration was an interesting technical achievement but not particularly useful to the target user group.

8.3 PLATFORMS

The choice of implementation platform for a video CODEC depends on a number of factors including the type of application, development support, power consumption restrictions, the need for future product upgrades and the availability and cost of commercial CODECs for the platform. Table 8.2 compares some popular implementation platforms (dedicated hardware, Digital Signal Processor or 'media processor', embedded processor or PC) and makes some general comments about their relative advantages and disadvantages Further discussion of platform capabilities can be found in reference [2].

At the present time, PC software and dedicated hardware are probably the most widely-used implementation platforms for MPEG-4 Visual. DSP and embedded platforms are becoming popular for mobile video applications because of their good balance between power efficiency, performance and flexibility.

8.4 CHOOSING A CODEC

Choosing a CODEC can be difficult, not least because each vendor presents the performance and capabilities of their CODEC in a different way. Issues to consider include availability and

Table 8.2 Platform comparisons

Platform	Advantages	Disadvantages
Dedicated hardware	Performance and power efficiency (best)	Inflexible, high development cost
DSP or media processor	Performance and power efficiency (good); flexibility	Limited choice of CODECs; medium development cost; single vendor
Embedded processor	Power efficiency (good); flexibility	Poor performance; single vendor
General purpose processor (e.g. PC)	Performance (medium/good); flexibility (best); wide choice of CODECs	Poor power efficiency

licensing terms, the support for required Profiles and Levels, subjective quality, compression performance, computational performance (e.g. instructions per second or coded frames per second), the suitability for the chosen target platform (e.g. whether it has been optimised for this specific platform) and the interface (API) to the CODEC. Ideally, the CODEC performance should be evaluated using source video material that is representative of the target application (rather than, or as well as, using 'standard' test video sequences).

Table 8.3 lists a selection of commercially-available MPEG-4 Visual CODECs, based on information available from the MPEG-4 Industry Forum web site (www.m4if.org) in March 2003. This is by no means a complete list of MPEG-4 CODECs but gives a flavour of the range of software and hardware implementations currently available. It is worth noting that most of the information in this table is based on manufacturers' claims and is not guaranteed to be correct.

The majority of the CODECs listed here appear to be targeted at streaming or storage applications. There are many software implementations of MPEG-4 Visual available, ranging from the official reference software (full-featured but far from real-time) to highly optimised real-time players. Hardware implementations tend to be targeted at high-performance (e.g. broadcast-quality encoding) or low-power (e.g. mobile streaming) applications. Several manufacturers supply code for a particular Digital Signal Processor (DSP) or Embedded Processor or System on Chip (SoC) modules suitable for integration into a larger hardware design or implementation on a Field Programmable Gate Array (FPGA).

The most common profile supported is Simple Profile, followed by Advanced Simple Profile (offering its more efficient coding performance). Two companies in this list (Dicas and Prodys) offer Core Profile solutions but of these two, only Dicas' CODEC supports the binary shape tool of the Core Profile. Significantly, no other profiles of MPEG-4 Visual are supported by any of these CODECs.

The list of commercial H.264 CODECs (Table 8.4) is shorter and details on some of these CODECs are very limited. This is not surprising since at the time of publication H.264 has only just been released as a full Recommendation/International Standard. Early adopters (such as those listed here) run the risk of having to change their designs to accommodate late modifications to the standard but have the potential to capture early market share.

Where information is available, Main Profile seems to be the most popular amongst these early implementations, possibly because of its support for interlaced video and more efficient high-latency video (through the use of B-slices). This implies that initial target applications

Table 8.3   MPEG-4 Visual CODECs (information not guaranteed to be correct)

Company	HW/SW	Profiles	Performance	Comments
Amphion www.amphion.com	HW	SP	L0-L3	SoC modules, plus HW accelerators
Dicas www.dicas.de	SW	SP, ASP, Core	Up to 2048 × 2048/60 fps?	Implements binary shape functionalities
DivX www.divx.com	SW	SP, ASP	All levels	Now compatible with MPEG-4 File Format
Emblaze www.emblaze.com	HW	SP	QCIF/up to 15 fps encode, 30 fps decode	Based on ARM920T core, suitable for mobile applications
EnQuad www.enquad.com	SW?	Core?	30 fps?	No product details available
Envivio www.envivio.com	HW/SW	SP and ASP	L0-L5	SW and HW versions
Equator www.equator.com	SW	SP	?	Decoder (running on Equator's BSP-15 processor)
Hantro www.hantro.com	HW/SW	SP	L0-L3	SW and HW versions
iVast www.ivast.com	SW	SP/ASP	L0-L3	
Prodys www.prodys.com	SW	SP/ASP/Core	L0-L4 (ASP)	Implemented on Texas Instruments TMS320c64× processor. Does not implement binary shape coding.
Sciworx www.sci-worx.com	HW/SW	SP	QCIF/15 fps (encoder)	Embedded processor solution (partitioned between hardware and software)
Toshiba www.toshiba.com/taec/	HW	SP	QCIF/15 fps encode + decode	Single chip including audio and multiplex
IndigoVision www. indigovision.com	HW	SP	L1-L3	SoC modules
3ivx www.3ivx.com	SW	SP/ASP	?	Embedded version of decoder available
UBVideo www.ubvideo.com	SW	SP	Up to L3	PC and DSP implementations

for H.264 will be broadcast-quality streamed or stored video, replacing existing technology in 'higher-end' applications such as broadcast television or video storage.

8.5 COMMERCIAL ISSUES

For a developer of a video communication product, there are a number of important commercial issues that must be taken into account in addition to the technical features and performance issues discussed in Chapters 5, 6 and 7.

Table 8.4 H.264 CODECs (information not guaranteed to be correct)

Company	HW/SW	Supports	Performance	Comments
VideoLocus www.videolocus.com	SW/HW	Main profile	30 fps/4CIF, up to level 3	HW/SW encoder; SW decoder; sub-8 × 8 motion compensation not yet supported
UBVideo www.ubvideo.com	SW	Main profile	30fps/4CIF	DSP implementation (Texas Instruments TMS320C64×)
Vanguard Software Solutions www.vsofts.com	SW	?	?	Downloadable Windows CODEC
Sand Video www.sandvideo.com	HW	Main profile	Supports high definition (1920 × 1080)	Decoder
HHI www.hhi.de	SW	Main profile	?	Not real-time yet?
Envivio www.envivio.com	SW/HW	Main profile	D1 resolution encoding and decoding in real time (HW)	Due to be released later in 2003
Equator www.equator.com	SW	?	?	Implementation for BSP-15 processor; no details available
DemoGraFX www.demografx.com/	SW/HW	?	?	Advance information: encoder and decoder will include optional proprietary 'extensions' to H.264
Polycom www.polycom.com	HW ?	?	?	Details not yet available
STMicroelectronics us.st.com	HW	?	?	Advance information: encoder and decoder running on Nomadik media processor platform
MainConcept www.mainconcept.com	SW	?	?	Advance information, few details available
Impact Labs Inc., www.impactlabs.com	SW	?	?	Advance information, few details available

8.5.1 Open Standards?

MPEG-4 and H.264 are 'open' international standards, i.e. any individual or organisation can purchase the standards documents from the ISO/IEC or ITU-T. This means that the standards have the potential to stimulate the development of innovative, competitive solutions conforming to open specifications. The documents specify exactly what is required for conformance, making it possible for anyone to develop a conforming encoder or decoder. At the same time, there is scope for a developer to optimise the encoder and/or the decoder, for example to provide enhanced visual quality or to take best advantage of a particular implementation platform.

There are however some factors that work against this apparent openness. Companies or organisations within the Experts Groups have the potential to influence the standardisation process and (in the case of MPEG) have privileged access to documents (such as draft versions of new standards) that may give them a significant market lead. The standards are not easily approachable by non-experts and this makes for a steep learning curve for newcomers to the field. Finally, there are tens of thousands of patents related to image and video coding and it is not considered possible to implement one of the more recent standards without potentially infringing patent rights. In the case of MPEG-4 Visual (and probably the Main and Extended Profiles of H.264), this means that any commercial implementation of the standard is subject to license fee payments (see below).

8.5.2 Licensing MPEG-4 Visual and H.264

Any implementation of MPEG-4 Visual will fall into the scope of a number of 'essential' patents. Licensing the rights to the main patents is coordinated by MPEG LA [1], a body that represents the interests of the major patent holders and is not part of MPEG or the ISO/IEC. Commercial implementation or usage of MPEG-4 Visual is subject to royalty payments (through MPEG LA) to 20 organisations that hold these patents. Royalty payments are charged depending on the nature of use and according to the number of encoders, decoders, subscribers and/or playbacks of coded video. Critics of the licensing scheme claim that the cost may inhibit the take-up of MPEG-4 by industry but supporters claim that it helpfully clarifies the complex intellectual property situation and ensures that there are no 'hidden costs' to implementers.

H.264/MPEG-4 Part 10 is also subject to a number of essential patents. However, in order to make the new standard as accessible as possible, the JVT has attempted to make the Baseline Profile (see Chapter 6) 'royalty free'. During the standardisation process, holders of key patents were encouraged to notify JVT of their patent claims and to state whether they would permit a royalty free license to the patent(s). These patent statements have been taken into account during the development of the Profiles with the aim of keeping the Baseline free of royalty payments. As this process is voluntary and relies on the correct identification of all relevant patents prior to standardisation, it is not yet clear whether the goal of a royalty-free Profile will be realised but initial indications are positive[1]. Implementation or use of the Main and Extended Profiles (see Chapter 6) is likely to be subject to royalty payments to patent holders.

8.5.3 Capturing the Market

Defining a technology in an International Standard does not guarantee that it will be a commercial success in the marketplace. The original target application of MPEG-1, the video CD, was not a success, although the standard is still widely used for storage of coded video in PC-based applications and on web sites. MPEG-2 is clearly a worldwide success in its applications to digital television broadcasting and DVD-Video storage. The first version of MPEG-4 Visual was published in 1999 but it is still not clear whether it will become a market leading technology for video coding. The slow process of agreeing licensing terms (not

---

[1] In March 2003, 31 companies involved in the H.264 development process and/or holding essential patents confirmed their support for a royalty-free Baseline Profile.

finalised until three years after the publication of the standard) and the nature of these terms has led some commentators to predict that MPEG-4 Visual will not 'capture' the market for its target applications. At the same time, it is widely agreed that there is a requirement to find a successor to the popular (but rather old) technology of MPEG-2. Some developers may opt for H.264 and bypass MPEG-4 Visual completely but despite its clear performance advantages (see Chapter 7), H.264 is the less mature of the two standards. There is a continuing debate about whether proprietary CODECs (for example, Microsoft's Windows Media Player 9 [3]) may offer a better solution than the standards. Developers of proprietary CODECs are not constrained by the tightly-regulated standardisation process and can (arguably) upgrade their technology and respond to industry requirements faster than standards-based vendors. However, reliance on a single supplier (or at least a single primary licensor) is seen by some as a significant disadvantage of proprietary solutions.

8.6 FUTURE DIRECTIONS

Guessing the future commercial and technical evolution of multimedia applications is a notoriously inexact science. For example, videoconferencing (now a mature technology by most standards) is yet to be as widely-adopted as has been predicted; many of the innovative tools that were introduced with the MPEG-4 Visual standard five years ago have yet to make any commercial impact. However, some predictions about the short, medium and long-term development of the technologies described in this book are presented here (many of which may be proved incorrect by the time this book is published!).

In the short term, publication of Recommendation H.264/MPEG-4 Part 10 is likely to be followed by amendment(s) to correct omissions and inaccuracies. A number of ongoing initiatives aim to standardise further transport and storage of MPEG-4 and H.264 coded video data (for example, file and optical disk storage and transport over IP networks). Following the extended (and arguably damaging) delay in agreeing licensing terms for MPEG-4 Visual, it is hoped that terms for H.264 will be finalised shortly after publication of the Standard (hopefully with confirmation of the royalty-free status of the Baseline profile). The ongoing debate about which of the current standard and proprietary coding technologies is the 'best' is likely to be resolved in the near future. The main contenders appear to be MPEG-4 Visual, H.264 and Microsoft's proprietary Windows Media 9 ('Corona') format [3].

In the medium term, expect to see new Profiles for H.264 adding support for further applications (for example, Studio and/or Digital Cinema coding) and improved coding performance (probably at the cost of increased processing requirements). For example, there is currently a call for proposals to extend the sample depth supported by the standard to more than eight bits per pixel and to include support for higher chrominance quality (4:2:2 and 4:4:4) [4]. It is difficult to see how MPEG-4 Visual can develop much further, now that the more efficient H.264 is available. The 'winner' of the coding technology debate will begin to replace MPEG-2 and H.263 in existing applications such as television broadcasting, home video, videoconferencing and video streaming. Some early mobile video services are based on MPEG-4 Visual but as these services become ubiquitous, H.264 may become a more favoured technology because of its better performance at low bitrates. Widespread adoption of mobile video is likely to lead to new, innovative applications and services as users begin to engage with the technology and adapt it to suit their lifestyles.

In the longer term, expect to see the emergence of new video coding standards as processor performance continues to improve and previously impractical algorithms become feasible in commercial applications. Video coding research continues to be extremely active and there are a number of promising approaches that may finally replace the long-standing DPCM/DCT coding paradigm. These include (among others) model-based, mesh-based and wavelet-based video compression. It has been argued that the DPCM/DCT model has had 15 years of continuous development and is near its performance limit, whereas other techniques offer the possibility of greater potential performance (however, this argument has been circulating for a number of years now and DPCM/DCT is not yet dead!) It is possible that some of the more esoteric features of MPEG-4 Visual (e.g. object-based and mesh-based coding) may re-emerge in future standards alongside improved coding technology. It could be said that these tools are ahead of their time at present and may have more chance of commercial success if and when real application needs become apparent.

8.7 CONCLUSIONS

Years of effort by many hundreds of researchers and developers have led to the standardisation of MPEG-4 Visual and H.264/MPEG-4 Part 10/AVC. The standards are impressive achievements, each in a different way. MPEG-4 Visual adopts an imaginative and far-sighted approach to video compression and many of its features and tools are perhaps still ahead of their time. H.264 has taken a more pragmatic, focused approach to addressing the problems and needs of current and emerging multimedia applications. This book has attempted to explain the fundamentals of both standards and to put them in the context of the goals of the standardisation groups and the ever-changing market. The answer to the much-asked question, 'which standard is best for my application', is not yet clear, although there are indications that H.264 may become the technical leader that will drive the next generation of digital video applications. However, before staking a reputation or a product development strategy on the likely outcome of the 'MPEG-4 vs. H.264' debate, it is worth remembering a similar debate from the distant past: VHS *vs*. Betamax.

8.8 REFERENCES

1. MPEG LA, http://www.mpegla.com.
2. Iain E G Richardson, 'Video Codec Design', John Wiley & Sons, 2002
3. Microsoft Windows Media 9 Series, http://www.microsoft.com/windows/windowsmedia/.
4. ISO/IEC JTC1/SC29/WG11 N5523, Call for Proposals for Extended Sample Bit Depth and Chroma Format Support in the Advanced Video Coding Standard, March 2003.

Bibliography

1. A. Puri and T. Chen (eds), *Multimedia Systems, Standards and Networks*, Marcel Dekker, 2000.
2. A. Sadka, *Compressed Video Communications*, John Wiley & Sons, 2002.
3. A. Walsh and M. Bourges-Sévenier (eds), *MPEG-4 Jump Start*, Prentice-Hall, 2002.
4. B. Haskell, A. Puri, A. Netravali, *Digital Video: An Introduction to MPEG-2*, Chapman & Hall, 1996.
5. F. Pereira and T. Ebrahimi (eds), *The MPEG-4 Book*, IMSC Press, 2002.
6. I. E. G. Richardson, *Video Codec Design*, John Wiley & Sons, 2002.
7. K. K. Parhi and T. Nishitani (eds), *Digital Signal Processing for Multimedia Systems*, Marcel Dekker, 1999.
8. K. R. Rao and J. J. Hwang, *Techniques and Standards for Image, Video and Audio Coding*, Prentice Hall, 1997.
9. M. Ghanbari, *Video Coding: An Introduction to Standard Codecs*, IEE Press, 1999.
10. M. J. Riley and I. E. G. Richardson, *Digital Video Communications*, Artech House, 1997.
11. V. Bhaskaran and K. Konstantinides, *Image and Video Compression Standards: Algorithms and Architectures*, Kluwer, 1997.
12. W. B. Pennebaker, J. L. Mitchell, C. Fogg and D. LeGall, *MPEG Digital Video Compression Standard*, Chapman & Hall, 1997.
13. IEEE Transactions on Circuits and Systems for Video Technology, special issue on H.264/AVC, to appear in 2003.

H.264 and MPEG-4 Video Compression: Video Coding for Next-generation Multimedia.
Iain E. G. Richardson.   © 2003 John Wiley & Sons, Ltd. ISBN: 0-470-84837-5

Index

H.264 and MPEG-4 Video Compression: Video Coding for Next-generation Multimedia.
Iain E. G. Richardson.   © 2003 John Wiley & Sons, Ltd. ISBN: 0-470-84837-5